Data Control

Smart Innovation Set

coordinated by
Dimitri Uzunidis

Volume 29

Data Control

Major Challenge for the Digital Society

Jean-Louis Monino

ISTE

WILEY

First published 2020 in Great Britain and the United States by ISTE Ltd and John Wiley & Sons, Inc.

ISTE Ltd
27-37 St George's Road
London SW19 4EU
UK

www.iste.co.uk

John Wiley & Sons, Inc.
111 River Street
Hoboken, NJ 07030
USA

www.wiley.com

Library of Congress Control Number: 2020938410

British Library Cataloguing-in-Publication Data
A CIP record for this book is available from the British Library
ISBN 978-1-78630-550-3

The manufacturer's authorized representative according to the EU General Product Safety Regulation is Wiley-VCH GmbH, Boschstr. 12, 69469 Weinheim, Germany, e-mail: Product_Safety@wiley.com.

Contents

Foreword

In the fight against waste, one particular commodity, however eminently precious, is rarely mentioned: data.

While it is not a priority, it is first collected to serve as, or illustrate, a reasoning or an idea. The major challenge of the last two decades has refocused on another field: that of understanding and analysis. Offering a clear report facilitates the management and steering of an activity.

Data processing opens the door to strategic recommendation. Now that the value of data is known, its value can further improve, provided it is operational and ready to use.

The individual, the consumer and the voter are regularly solicited, sometimes against their will, often in ignorance of the risks involved. In the face of the abuses that have been noted, the law has had to be involved.

Before this stage of maturity, "just in case" storage is reassuring without really knowing why. This intuition leads companies to sometimes accumulate incomplete or unorganized data. Ignorance of its purpose has resulted in the accumulation of data in huge volumes, with the only limit being the technical constraints of the moment, the cost per byte.

So far, it has been impossible to use all this data due to lack of time and resources. Therefore, it is not possible to educate the overall population. Using a sample is imperative, with representativeness being a key concept.

Technological progress has made it possible to overcome this constraint of representativeness. The entire dataset is now vulnerable to attack even though the size is discouraging. These masses of data, veritable "assembled pieces", have whetted the appetite of the greedy known as Big Data, machine learning, artificial intelligence and other "datavores", reinventing a 2.0 statistic: data science.

At the same time, the proliferation of equipment (tools, applications) has democratized the use of data, making connection and leading to automatic and endemic branches. The variety of media (computers, tablets, smartphones), the explosion of networks and access providers facilitate the exchange, sharing and processing of information. Mobile consumption has become instantaneous.

This continuous feeding creates a new addiction, a demand of a different nature. In order to respond before it becomes obsolete, it is necessary to move quickly, even if it means returning the information to its original role as simple data. In this sense, pre-selections are made, and the data is delivered, sorted and "ready to read", not necessarily ready to use. The preferences and tastes of the "target" are identified; its supposedly known expectations are identified in relation to its history. Trespassing?

However, the absence of filters or segmentation leads to fears of saturation. The data is ubiquitous despite its limited lifespan. It is refreshed and constantly updated. The required reactivity sometimes weighs on its quality. The corollary concession is the "average" information, which is difficult to exploit. It is against this risk of drowning or submersion that analysts, true lifeguards, fight. Their responsibilities are cleaning, reducing and synthesizing data; detailing or refocusing it if necessary; and using it in the delegated question. At the heart of the process, this human presence guarantees the surpassing of a simple mechanical treatment and therefore the creation of value. The mission of the latter consists, after all the stages of data processing, of restoring the true dimension of the data while making it profitable.

Here, Professor Monino's book brings the main tools that are indispensable to his mission, which, in the course of time, will become more and more indispensable. Unfortunately, not all companies have the necessary resources (time, budget, analyst) to extract or exploit this wealth. However, the use of specialists such as E2S-Conseils allows us to make the most of this information to help companies develop and flourish.

I hope that you enjoy reading and learning from this book.

Philippe RIBIERE
Chief Executive Officer
E2S-Conseils

Acknowledgements

I wish to dedicate this book to my sister who unfortunately was unable to read my work on data.

Also, all the members of the "*Réseau de Recherche sur l'Innovation*", RRI, and in particular, Dimitri Uzunidis, who encouraged me to publish this work and I thank him warmly for his encouragement.

This book is the culmination of many years of research dedicated to data processing, to statistics and to econometrics in the TRIS laboratory (*Traitement et Recherche de l'Information et de la Statistique*). It is the fruit of various work carried out within the framework of R&D (Research & Development) for several start-ups in the Languedoc-Roussillon of France and large private and public groups.

Thank you to all those who have supported me during difficult times and have helped transform an individual intellectual adventure into a collective one. In particular I offer my thanks to Philippe RIBIERE, President of E2S-Conseils with whom I have written several R&D books over the years.

Thank you also to the team of researchers at the "*Montpellier Recherche en Economie*" laboratory who demonstrated their trust in my research and granted me my emeritus status.

I would also like to thank my children: Christine, Laurent, Caroline and Daniel, as well as their partners: Laurent, Alexandra, Guillaume and Louise who have all supported and surrounded me during the difficult times I had to face.

To my grandchildren: Lily-Rose, Jean-Baptiste, Alice and Gabriel.

To Anne-Marie, my wife.

Jean-Louis MONINO

Introduction

The world has become digital, and technological advances have increased the number of different ways for accessing, processing, and distributing data. New technologies are now reaching a certain maturity.

Today, data comes from all sides: geolocation sensors, smartphones, social networks where we share files, videos, photos, etc., Internet shopping transactions by customers, banking transactions through credit cards and so on. In France, out of 65 million people, 83% are Internet users; 42% are registered on Facebook, or 28 million members. More than 72 million phones are activated, and the French spend on average more than 4 hours a day surfing the Internet. French mobile users spend 58 minutes or more there; 68% of the population is registered on social networks. French people spend more than 1 hour and 30 minutes a day on social networks. The development of these masses of data and their access represents what is called "Big Data". These immaterial data arrive continuously; their processing poses problems, particularly in knowledge extraction. Thus, new methods of automatic information extraction are implemented: for example, "data mining" or "text mining". They underlie profound changes that affect the economy, marketing, research and even politics. The amount of data will increase sharply with the arrival on the market of connected objects that will gradually come into use. Elements of our daily life are already connected: the car, the television and some household appliances. They are or will be equipped with a chip to collect and transmit data to their users via a computer, tablet or smartphone. The most important thing is that these items will also be able to trade with each other! This will allow us to remotely manipulate the equipment in our home, in our vehicles by connecting to them from our home and outside our home, using smartphones or any other equipment. This is called “the Internet of Things”.

This phenomenon is now of interest to operational decision-makers (marketing managers, financial managers, etc.) when it comes to analyzing the immense potential of data held by companies in real-time. In order to meet the challenge of “Big Data”, measures should be taken that include all the tools that allow data to be processed in a more restrictive way, as well as all the actors who analyze these data. This will only be possible through an awareness of the gains offered by “data enhancement”. The data in the organized, reorganized bases, processed by statistical methods or econometric modeling, become knowledge.

For a company it is essential to have more and more data about the environment in which it operates or will operate. We will no longer work on classes of behaviors but on individual analysis. It is easy to understand that this revolution is leading to the creation of so-called “start-up” companies whose aim is the automatic processing of data that make up what is known as “Big Data”. This is certainly one of the components of what some call the “new industrial revolution”. The Internet, digital and connected objects have opened up new horizons in a multitude of fields.

As an example, access to data allows quantitative and qualitative analyses to be enriched. Customer contacts can be analyzed with data collected by a call center[1]; this kind of product can also be offered in limited numbers, as E2S-Conseils does. This needs to be developed by exploring the content of emails, voice calls, and mixing this information with website navigation. Or, study the messages exchanged on social networks (Facebook, Twitter, Linkedin, etc.) to identify interesting new products or to know which products are most talked about.

For data to reach its full value potential, it must be easily accessible to all interested parties without additional barriers and at economically bearable costs. If the data are open to users (Matouk – MTRL review), other specific data processing companies may be created. This activity will meet the needs of users without the need for them to build models and equations themselves.

1 A call center is a center for processing incoming and/or outgoing calls.

In addition to its potential in economic terms and for the creation of new activities, the opening up of data is also a matter of philosophical or ethical choice. It quantifies collective human behavior and, therefore, belongs, in fact, to those whose behavior has been measured. The culture of this phenomenon is based on the availability of data towards a communication orientation.

To limit itself to the economic benefits of open data, a study carried out for the European Commission estimated the total public sector information market in 2008 in the EU at 28 billion euros. According to this study, the overall economic benefits of more open public sector data would amount to about 40 billion euros per year for the EU. For the EU economy as a whole, the total direct and indirect economic gains from the use of ISPs (Internet Service Providers) and applications based on these data would be in the order of 140 billion euros per year.

The Internet age has massively increased the search for information. Companies are overwhelmed by the flood of data that results from simply surfing the Internet. In other words, they are forced to acquire the relevant information to develop their high value-added strategies in order to master the ever-changing environment. Now, the management of industrial strategies is largely based on the ability of companies to access strategic information to better develop their environment. This information can therefore be the source of new knowledge (knowledge pyramid).

The process of collecting, processing and interpreting information is a practice that goes beyond the definition of ideas to the realization of those ideas to ensure a better production of knowledge that generates innovations. It is economic intelligence that enables each company to optimize its offer quantitatively and qualitatively and also to optimize its production technologies.

In addition to the advent of ICT, the speed of production, dissemination and data processing capacity, another element has become important in recent years: time. This temporal element implies a notion of speed of information flow. This calls for a rethinking of corporate strategy, beyond the difficulties posed by the processing of large volumes of data. The value of a data item increases over time and depends on the multiple uses that are made of it.

Remuneration in the digital professions rose sharply in 2018, according to Robert Half's remuneration study[2]. However, HR directors admit that it is often difficult to find the rare pearl: the average recruitment time for an executive manager is four months.

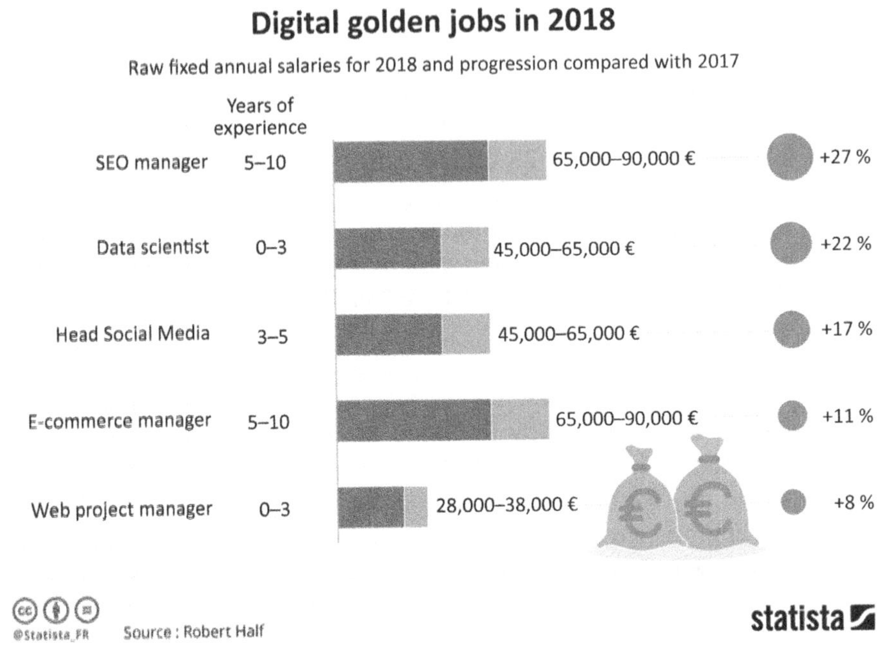

Figure I.1. *Digital golden jobs in 2018. For a color version of this figure, see www.iste.co.uk/monino/control.zip*

For some professions, the jump will be very high. According to Robert Half, the "golden job" in the digital world will be that of SEO Manager with 5–10 years of experience: the remuneration linked to this job should increase by 27% compared to 2017 and reach between 65,000 and 90,000 euros per year. Data scientists, even with little experience, will also see their salaries increase by 22%.

2 Robert Half's study: "Les emplois en or du numérique en 2018" by Claire Jenik, Statistica, 11 Oct. 2017.

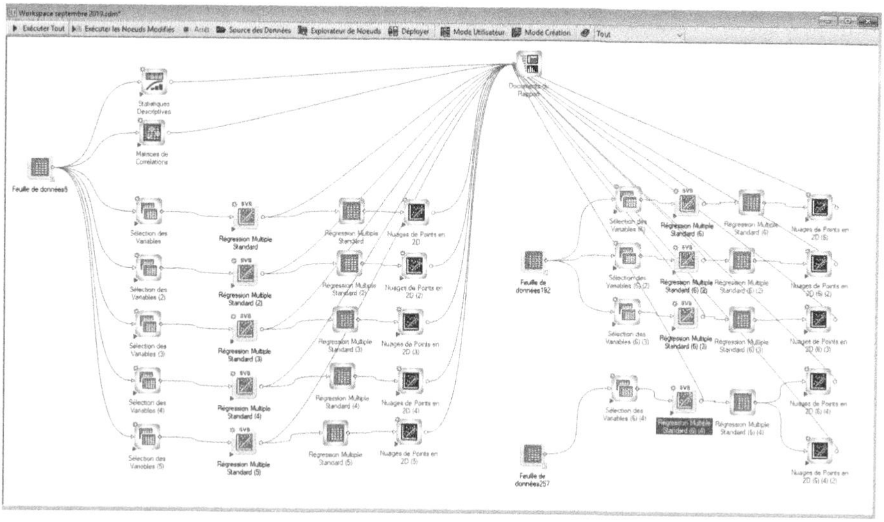

Figure I.2. *TIBCO's data mining and Statistica software. The research and development project of E2S-Conseils in collaboration with the TRIS laboratory has been developed using Tibco's Statistica Data Mining for health research (MONINO L., TIBCO's data mining and Statistica software. The research and development project of E2S-Conseils in collaboration with the TRIS laboratory has been developed using Tibco's Statistica Data Mining for health research, Thesis, 2019). For a color version of this figure, see www.iste.co.uk/monino/control.zip*

Big Data includes the processing of these large masses of data, from their collection and storage to their visualization and analysis. This data thus becomes the fuel of the digital economy. They constitute the raw material indispensable to the queen activity of the new century: "data intelligence". This book shows that the stakes of data are focused on the integration and development in the company. It deals with the issue of data control and valuation in a competitive context.

Faced with this multiplication of data, companies have to mobilize sophisticated processing techniques. In fact, the mastery of processing techniques is now becoming a real strategic and useful question for the competitive differentiation of companies (Bughin and Chui 2010). The processing of these masses of data plays a key role in tomorrow's society, with applications in fields as diverse as science, marketing, customer services, sustainable development, transportation, health and even education.

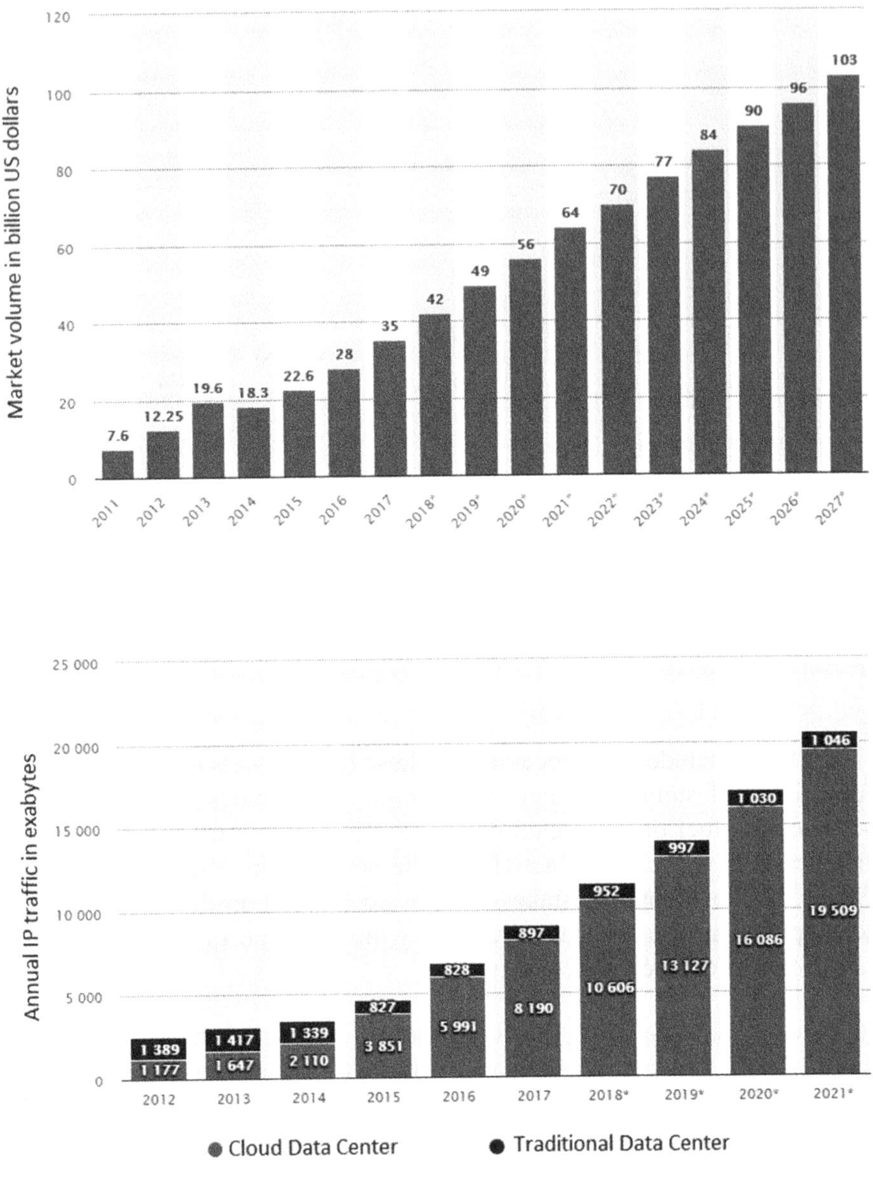

Figure I.3. *Forecasting Big Data market size, based on revenues, 2011–2027 (source: Statista). For a color version of this figure, see www.iste.co.uk/monino/control.zip*

In this context, companies must have the capacity to absorb all available data, enabling them to assimilate and reproduce knowledge. This capacity presupposes the existence of specific skills that enable the use of this knowledge. The training of "data scientists" is therefore essential in order to be able to identify useful approaches to opening up or internal exploitation of data and to quantify the benefits in terms of innovation and competitiveness, since Big Data is only one element of a new set of tools and techniques called "data science".

The Data Scientist's mission is to extract knowledge from company data. They will be called upon to perform strategic functions within the Commission. To do so, they must master the necessary tools. They must also be more pedagogical and increase their command of data mining, because the volume of data requires an increase in the range of techniques to be mastered.

Big Data includes the processing of these large masses of data, their collection, storage, visualization and analysis. This data thus becomes the driving force of the digital economy. They constitute the raw material indispensable to the queen activity of the new century: "data intelligence". This book shows that the stakes of Big Data are focused on the integration and the development of data in the company. It deals with the issue of data valuation in a context of strong competition.

More precisely, it represents a research subject that involves several fields (Big Data, open data, data processing, innovation, economic intelligence, etc.). This multidisciplinarity allows us to bring a considerable enrichment to studies and research on the development of data warehouses in its entirety.

Information on the French population was published by various news media in 2019, for example, that published by Paul Manuel Godoy Hilario on 16 May 2019. This statistic displays an interest in the French population in the new data through the different information media in 2019. Approximately half of the respondents said they were quite interested in this news. In 2018, nearly 35% of French people turned first to generalist television channels to further develop information they had received.

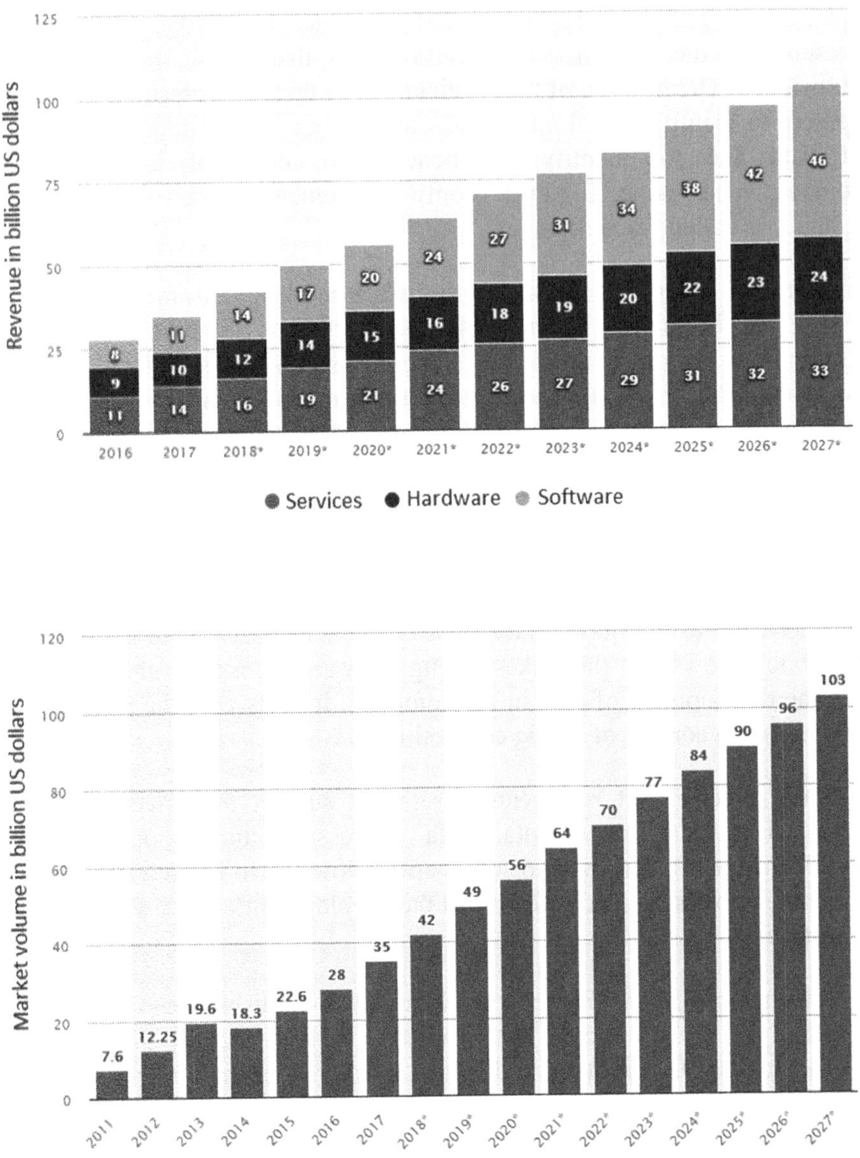

Figure I.4. *The Big Data turnover and market size (source: Statista). For a color version of this figure, see www.iste.co.uk/monino/control.zip*

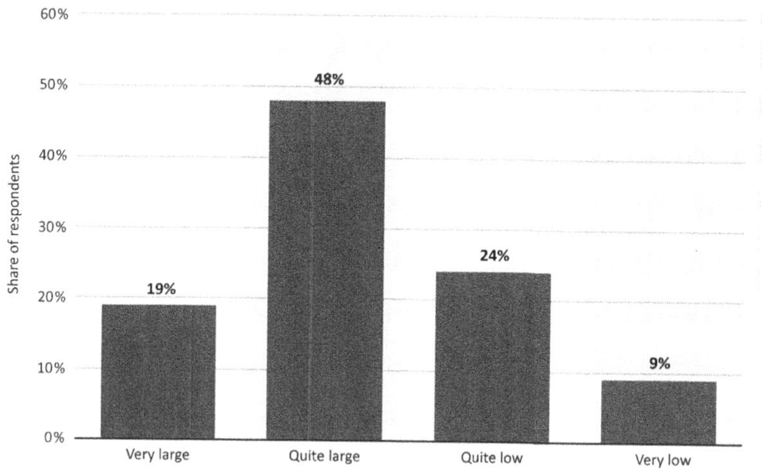

Figure I.5. *With what interest do you follow the news through the information media (press, radio, television)? (source: Statista). For a color version of this figure, see www.iste.co.uk/monino/control.zip*

This book is part of the continuation of the book *Big Data, Open Data and Data Development* published by ISTE and Wiley in their *Smart Innovation* set of books – coordinated by Dimitri Uzunidis – Innovation Research Network (Monino and Sedkaoui 2016). This book brings together all of my research work on economic intelligence over the last 20 years. Its objective is to show that the stakes of the "data revolution" era are focused on the integration and enhancement of data in the enterprise. It is part of a theme related to the rise of the intangible economy mobilizing knowledge and know-how, and highlights the importance of data. The contribution of information allows the generation of knowledge useful for decision-making, within the framework of the various activities specific to the development of the company. To do this, we must clarify what is more precisely a "Data" to be able to understand and formulate the expression "Big Data" issues.

We are going to follow this conception of EI (economic intelligence) (Monino and Boussetta 2013) step by step, based on the basic model proposed in 2005 by Monino and Lucato during the "mornings of the city" meetings at the Montpellier Chamber of Commerce and Industry. Data, information, knowledge and wisdom (decision) are all essential elements in preparing for a good decision and we will follow this in the following chapters of this book.

1

From Data to Decision-Making: A Major Pathway

While information is at the heart of economic intelligence, data are essential elements required to build knowledge in order to make a good decision. This is something which at the moment may seem optimal in the possible fields of its knowledge. It is the use of data that gives power. As companies become increasingly aware of the importance of data and information, they are rushing to think about how to "manage", enrich and leverage it.

Thus, the explosion of a phenomenal amount of data, and the need to analyze them, brings to the fore the well-known hierarchical model: "Data, Information and Knowledge". This model is often exploited in the literature on information and knowledge management. Several studies claim that the first appearance of the hierarchy of knowledge can be found in T.S. Elliot's poem "The Rock" in 1934. In recent literature, several authors cite R.L. Ackoff's 1989 publication "From data to wisdom" as a source of the hierarchy of knowledge. Indeed, this hierarchical model highlights three words: "Data", "Information" and "Knowledge". The relationship between these three words can be represented in the above form where knowledge is given the highest place to emphasize the fact that a great deal of data is necessary for the acquisition of knowledge.

This hierarchical model is often exploited in the literature on information and knowledge management. It can also be exploited as an approach to the concept of business intelligence.

1.1. Background on economic intelligence

In the United States, it was the work of academics in the 1960s that revealed the importance and necessity of conceiving economic intelligence as a branch of the economy. Harold Wilensky's book "Organizational Intelligence" – 1967. It defines business intelligence as the activity of producing knowledge serving the economic and strategic goals of an organization, collected and produced in a legal context and from open sources.

Stevan Dedijer in the late 1960s conceptualized "intelligence" as an economic matter, and gave a broad definition: "Intelligence is the information itself, and its processing, and the organization that deals with it, while it obtains, evaluates and uses it under more or less secret, competitive or cooperative conditions, for the purposes of conducting any social system and about the nature, capacities, intentions, actual or potential operations, of internal or external opponents."

Klaus Knorr was one of the first to advocate a wide dissemination of Business Intelligence, starting from the university space; for him it is "the operation to obtain and process information about the external environment in which an organization wants to maximize the achievement of its different goals". Business intelligence is becoming a major component of corporate strategy and is based on information that is the foundation of any decision-making process. Information as such had received a fundamental scientific treatment without which economic intelligence could not have developed.

In France, the contribution of a group of experts from the *Commissariat Général du Plan* was published in 1994 under the name of the Martre Report. This work on the "Economic Intelligence and Company Strategy" proposes a definition in the introduction to the report: "Economic intelligence can be defined as all the coordinated actions of research, processing and distribution, with a view to its exploitation, of information useful to economic players. These various actions are carried out legally with all the necessary guarantees of protection to preserve the company's assets, in the best conditions of quality, time and cost. Useful information is that which is needed by the various levels of decision-making in the company or community to develop and consistently implement the strategy and tactics necessary to achieve the objectives defined by the company in order to improve its position in its competitive environment. These actions, within the company, are organized in an uninterrupted cycle, generating a shared vision of the objectives to be achieved...."

1.2. Strategic economic intelligence revisited

The economy has become global, it is constantly changing. This dynamic gives information, which is at the heart of economic intelligence, a speed of circulation that can lead to radical changes in the economic environment of companies. Whoever masters or attempts to master information can have a significant edge in our competitive world. It also requires the ability to understand our economic, social and financial environment and its interactions very quickly. "Economic intelligence" is defined by ethical means, and its strength lies in the ability to interpret and not in the seditious nature of the information gathered. Economic intelligence is a major component of any strategy.

Economic Intelligence has had difficulty establishing itself in France. Nine years after the Martre report, the government asked MP Bernard Carayon for a new report. Five years after this last report, the 2008 Defense White Paper paid particular attention to economic intelligence in the context of the protection of fragile industrial sectors whose skills may be of a sensitive nature. Finally, in its progress report of January 5, 2010, the "*états généraux de l'industrie*" (general industry report) underlined the need for French industry to better understand its competitive environment, its potential markets and the opportunities that new technologies can offer. This report insists on support for exports through a better match between production and world demand through economic intelligence and a sustained promotion of "made in France".

1.2.1. *The three major steps for decision support*

The proposed model retains three concepts; "Data, Information, Knowledge" which make it possible to define the global concept of economic intelligence by highlighting the central place of "Information". This global concept leads to the decision-making and the cycle of Strategic Business Intelligence. This work was presented for the first time on 19 April 2005 at the CCI of Montpellier during "*Des Matins De La Cité*".

The explosion of a phenomenal amount of data, the need to analyze and visualize it, brings to the forefront the well-known hierarchical model: "Data, Information and Knowledge". This model is often exploited in the literature on information and knowledge management. Several studies claim that the first appearance of the hierarchy of knowledge can be found in T.S. Elliot's poem "The Rock" in 1934. The poem contained the following lines:

– Where is the wisdom we have lost in knowledge?

– Where is the knowledge we have lost in information?

In recent literature, several authors cite R.L. Ackoff's 1989 publication "From data to wisdom" as a source of the hierarchy of knowledge. Indeed, this hierarchical model highlights three words: "Data", "Information", "Knowledge"[1]. The relationship between these three words can be represented in the schematic form below, where knowledge takes the highest place to emphasize the fact that a lot of data is needed to acquire knowledge.

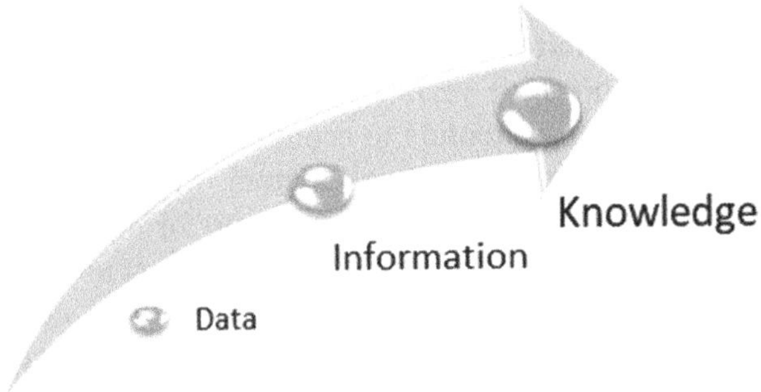

Figure 1.1. *The three concepts (source: Monino and Lucato 2005)*

1.2.2. *Modeling the concept of strategic business intelligence*

A schematization of the concept of Strategic Economic Intelligence can be proposed from part of the hierarchical model described by Thomas Stearns Eliot in 1934, which establishes a link between wisdom, knowledge

1 This hierarchical model was taken up in 2005 by J.-L. Monino and G. Lucato to show the importance of Information in the Business Intelligence process (Monino 2013).

and information starting from data. This model leads to wisdom (Ermine *et al.* 2012) that can be approached through strategic decision-making.

"I want the right information at the right time to make the right decision[2]." (Porter 1979).

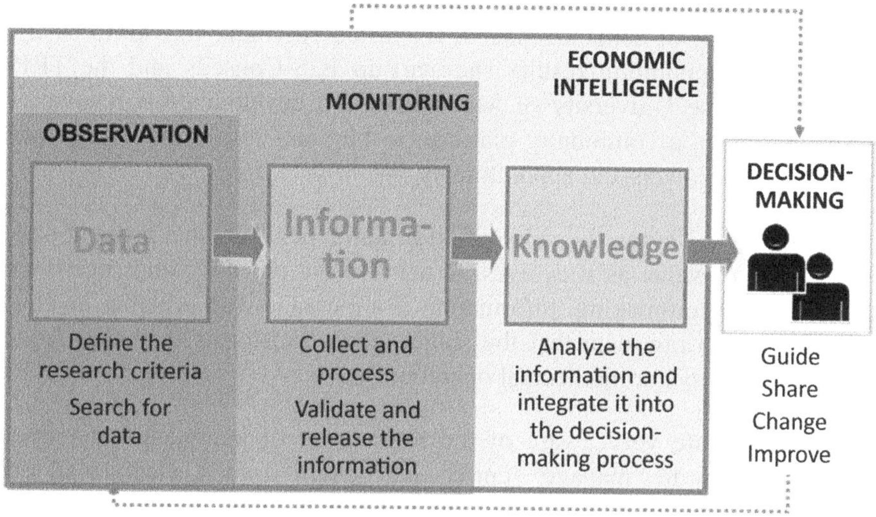

Figure 1.2. *Business intelligence model of data to decision-making (source: Monino and Lucato 2005). For a color version of this figure, see www.iste.co.uk/monino/control.zip*

1.2.2.1. *Data*

For a company, data are essential, and there are more and more concerning the environment in which it operates or will operate. We will no longer work on classes of behaviors, but on individual analysis. It is easy to understand that this revolution is leading to the creation of so-called "start-up" companies whose aim would be to automatically process the wealth of data that make up what is known as "Big Data". This is certainly one of the components of what some people call the new industrial revolution. The Internet, digital technology and connected objects have opened up new horizons in a multitude of fields.

2 Michael Porter: business intelligence is "giving the right information at the right time to make the right decision".

As an example, access to data allows quantitative and qualitative analyses to be enriched. We can analyze customer contacts with data collected by a call center, as well as offer this kind of product in limited numbers as E2S-Conseils does. Indeed, the web offers immense information possibilities, Internet users demand clear, precise and immediate information. The responsiveness of the demand for information is a key point in this type of demand. Indeed, it is often necessary to wait to have a more than disappointing result. The start-up E2S-Conseils and the TRIS laboratory of the University of Montpellier are developing an innovative service based on a community platform to buy and sell multi-product and multi-country contacts and appointments.

According to Taylor, the value of information begins with a piece of data, which acquires value as it evolves, to achieve the objective and specify an action for decision-making. Information is a message with a higher level of meaning. It is a raw data that the subject then transforms into knowledge through a cognitive or intellectual operation.

Companies are very aware of the importance of knowledge and even more so of how to "manage", enrich and benefit from it. Beyond all the factors (financial, technical and other), the knowledge that the company possesses also represents an important survival factor. Whether it is market knowledge, legal, technological, normative or other information.

This implies that in the information cycle, the data collected should be optimized so that needs are identified instantly and processed as quickly as possible. This will highlight the interaction between a multitude of actors (decision-makers, analysts, advisers, technicians, etc.) within the framework of a group dynamic and will allow the complementarity of knowledge, to improve understanding, analysis of situations, and production of information necessary for action. Indeed, "The quality of the production of knowledge for operational purposes depends on the skills of interpretation and analysis of the human factor in collective problem-solving situations" (Bulinge 2006, 2007).

1.2.2.2. *From data to information: monitoring*

Business intelligence concerns data collection, whereas business intelligence is a request from the company's management to know its environment. Of course, economic intelligence must mobilize human and financial means, but it also calls upon working methods, computerized or not, tools for collecting and especially analyzing these data. The data once collected, verified, processed and validated becomes strategic information of the company. The company manager must be able to find the necessary support when the decision is taken. This procedure makes it possible to consider:

– a set of possibilities;

– execution constraints;

– simulations to calculate indicators of success and failure.

In her research in 2008, Lesca addresses the problem of interpreting data to transform it into strategic information or knowledge. Interpretation systems, the heart of the "business intelligence" process, are defined as the set of "systems for attributing meaning" to the information that the company receives, manipulates and preserves (Baumard and Benvenuti 1998).

1.2.2.3. *From intelligence to economic intelligence*

"Monitoring" involves gathering, storing and sharing information according to identified needs and predefined procedures. In our concept, monitoring brings data together and transforms it into information and its transformation into useful data.

"Economic Intelligence" allows the analysis of the information obtained and prospective thinking to anticipate changes and make decisions in accordance with the economic environment. But it is also the perception of its environment, the identification of market developments, strategic adaptation, innovation and sales growth, the opening of new markets and the development of the company. Finally, it is information security, data protection, risk anticipation and, therefore, the control of its information system.

All the fields that complement Business Intelligence, such as knowledge management, information protection and lobbying, can be grouped together in the overall concept of Strategic Intelligence.

1.2.2.4. *Information*

Information is data that has been recorded, classified, organized, linked and interpreted within the context of a specific study. The exploitation of the collected data requires sorting, verification, processing, analysis and synthesis. It is through this process that the raw data collected during the research will be transformed into information. This data processing process provides accessible information for decision-making.

1.2.2.5. *From information to knowledge: economic intelligence*

Information is at the heart of Business Intelligence (Monino 2012b) and its strategic management has become one of the key drivers of overall business performance. Indeed, the process of globalization of markets forces economic agents to adapt to the new balances that are being established between competition and cooperation. From now on, the management of industrial strategies is largely based on the ability of companies to access strategic information to better anticipate future markets and competitors' strategies.

1.2.2.6. *Knowledge*

Knowledge is the extension of information, to which value is added because it is based on intention. It is a specifically human phenomenon that concerns thinking in its active and contextual relational frameworks. It represents an acquisition of information that is translated by the human factor and requires analysis and synthesis to assimilate, integrate, criticize and admit new knowledge.

1.2.2.7. *From knowledge to decision-making: strategic business intelligence*

Economic foresight concerns the collection of information, while competitive intelligence is a request from the company's management to know, and control as far as possible, the dynamics of its environment. Of course, economic intelligence must not only mobilize means, both human and financial, but it also requires working methods and tools, computerized or not, for the collection and especially the analysis of this information. Information thus becomes one of the strategic raw materials of the company and the person in charge of a company must be able to find the necessary help within his staff or externally when taking strategic decisions.

The wave of computerization that has swept over our societies is now reaching a form of maturity, which means that entry into the knowledge society represents a plunge into an ocean of information. The development of related technologies, with the advent of the Internet and networks, has increased the volumes available.

The stakes in the era of the "data revolution" are focused on the integration and enhancement of data. This phenomenon is part of a new approach to the "economy of the immaterial". It mobilizes knowledge and information, and highlights the importance of "Data". "Big Data", "Open Data", [...] "Data" is everywhere – we produce it every day – and is carried by smartphones, tablets, [...] and is accessible by anything and everyone. Its mass arrival is upsetting traditional management practices.

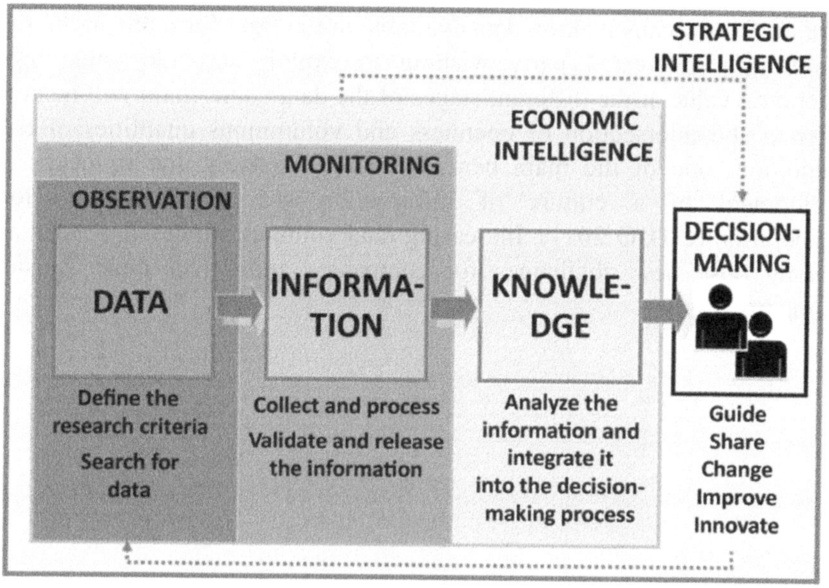

Figure 1.3. *From data to decision-making strategic intelligence (source: Monino and Lucato 2005, Conference Fondation Maison du Maroc Paris October 2019 "Economic intelligence and data control in the digital world"). For a color version of this figure, see www.iste.co.uk/monino/control.zip*

1.3. Conclusion

The advent of "Big Data" (Monino 2015) is changing traditional management practices and invites the organization to adopt new visions that allow it to have "strategic information" that can generate more value. In other words, the message must be conveyed that information per se is not power: it is using information that gives power, and the more information is given and exchanged, the more information is received (Martinet and Marti 2001).

It is then necessary to adapt and try new approaches, new methods, new knowledge and new ways of working, which leads to new properties and new issues since a logical referencing must be created and implemented. The question is no longer what data to store, but what can be done with that data?

However, reaching the full potential of data depends on how it is presented, it must be used and reused in multiple ways without diminishing its value. This means making data available in the right form and at the right time to any interested party wishing to exploit and add value to it. Generating value at the different stages of the data value chain will be at the center of the intersection of openness and voluminous quantities of data. In addition, one of the main benefits of open data is that it fosters the development of a culture of information sharing and intersectoral collaboration (GTDO 2011). Increasing data volumes, processing speed and diversity pose new challenges for companies and affect their decision-making processes.

2

Data: An Indispensable Platform for Companies

"Data is the new oil"

(Rotella 2012).

They constitute the raw material indispensable to the trademark activity of the new century: data intelligence. But we must be careful because a lot of data is not always the "right data". This is therefore an underlying issue that does not mean that the quality in the dataset is necessarily better. Good results can be achieved by making better use of available data. The latest technologies such as Smart, 4G and Cloud, as well as the advent of the Internet, data processing and data mining software have enabled the exploitation of data.

Today, data comes to us from everywhere: geolocation sensors, smartphones and social networks, where we share files, videos, photos of customers' online shopping transactions and banking transactions via bank cards.

Data has always had strategic value. According to Taylor (1980), the value of information begins with an input that acquires value as it evolves to achieve the objective of expressing a decision-making action. But with the breadth of data available today and the capacity to process it, it has become a new asset class.

The acceleration of the digital age and technological advances have turned information scarcity into information overload. The development of these masses of data and access to them represents what is called "Big Data". This mass of immaterial data that arrives in a continuous flow and its processing pose problems especially in knowledge extraction. The challenges around data are many: how to collect, store, analyze, disseminate and protect it?

Thus, new methods of automatic data extraction are being introduced, such as "data mining" or "text mining" (Monino and Boya 2011), which underpin profound changes, affecting both the economy and marketing, but also research. The amount of data will increase sharply with the arrival on the market of connected objects that we will use. Many everyday objects are already connected, such as the car, the television and some household appliances. They are or will be equipped with a chip to collect and transmit data to their users through a computer, a tablet, a smartphone and so on. The most important thing is that these objects will also be able to exchange with each other! We will thus be able to remotely manipulate the equipment in our home and in our vehicle by connecting either at home or outside our home using smartphones or any other equipment. This is what we call "the Internet of Things".

More and more connected objects are measuring gigantic amounts of data for forecasting, analysis or prediction purposes. Connected objects will indeed collect much more data on individuals, and the number of usable and monetizable data will increase.

The General Data Protection Regulation (GDPR) is the new European Union legal framework governing the collection and processing of users' personal data. It applies to all entities established in a European country that process personal data, as well as to all entities around the world that process personal data belonging to EU residents.

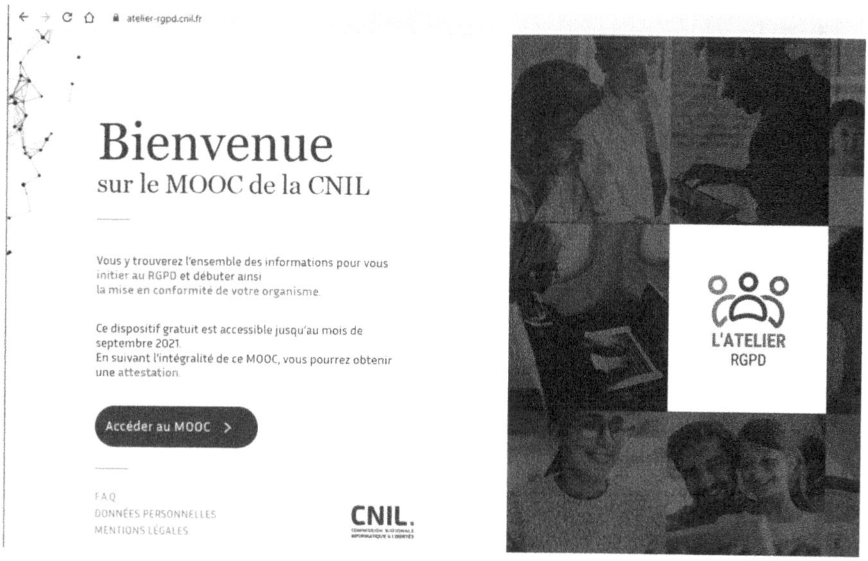

Figure 2.1. *MOOC[1] introduces you to the GDPR[2]. The CNIL offers professionals the opportunity to discover or better understand the GDPR, in order to enable them to initiate compliance within their organization and help raise awareness among operational staff[3]. For a color version of this figure, see www.iste.co.uk/monino/control.zip*

The ANSSI[4] has developed a set of information to introduce you to cybersecurity, deepen your knowledge, and thus act effectively on the protection of your digital tools.

The data is accessible to all and everywhere in the world, and the number of Internet users in 2018 was 4.4 billion or 57% of the world's population. There is a need for knowledge in order to apprehend these masses of data. We must inform, train and educate en masse.

---

1 A MOOC is an open type of distance learning that can accommodate a large number of participants. The term MOOC has become part of everyday language in France and is now recognized by the main dictionaries (Wikipedia).

2 GDPR: the General Data Protection Regulation is the new European Union legal framework governing the collection and processing of users' personal data. It entered into force on May 25, 2018.

3 Third French-speaking MOOC in number of active users (38,000 in 3 months). Developed by CNIL lawyers and experts – Hosting the device on social networks – Hosted via OnlineManager https://www.onlineformapro.com/cnil_atelier_rgpd/.

4 The ANSSI is the national authority for the security and defense of information systems. Prevention, protection, reaction, training, etc.

Figure 2.2. *ANSSI and information security. Raising French people's awareness of digital security is a major challenge. In response, ANSSI is launching its first online course, the MOOC SecNumacadémie. For a color version of this figure, see www.iste.co.uk/monino/control.zip*

2.1. The key figures of digital technology

When faced with a dataset, researchers need to understand not only the limitations of the available dataset, but also the limitations of the questions that the dataset must answer, and what interpretations are appropriate.

> The international agency We Are Social and the social networking platform Hootsuite published their Digital Report 2019, which delivered the latest trends in digital, social media, mobile, and e-commerce around the world.
>
> Hootsuite and We Are Social's new Digital Report presented the latest international data and trends in digital, social media, mobile, and e-commerce as of July 2019, to better understand the behavior of Internet users to date.
>
> Hootsuite is a social networking tool created by Ryan Holmes in 2008
>
> The complete study is available in the SlideShare and we will present some of the results of this study to get a vision of the place of digital, and the Internet in France and in the world.

Box 2.1. *The platform of the international agency We Are Social*

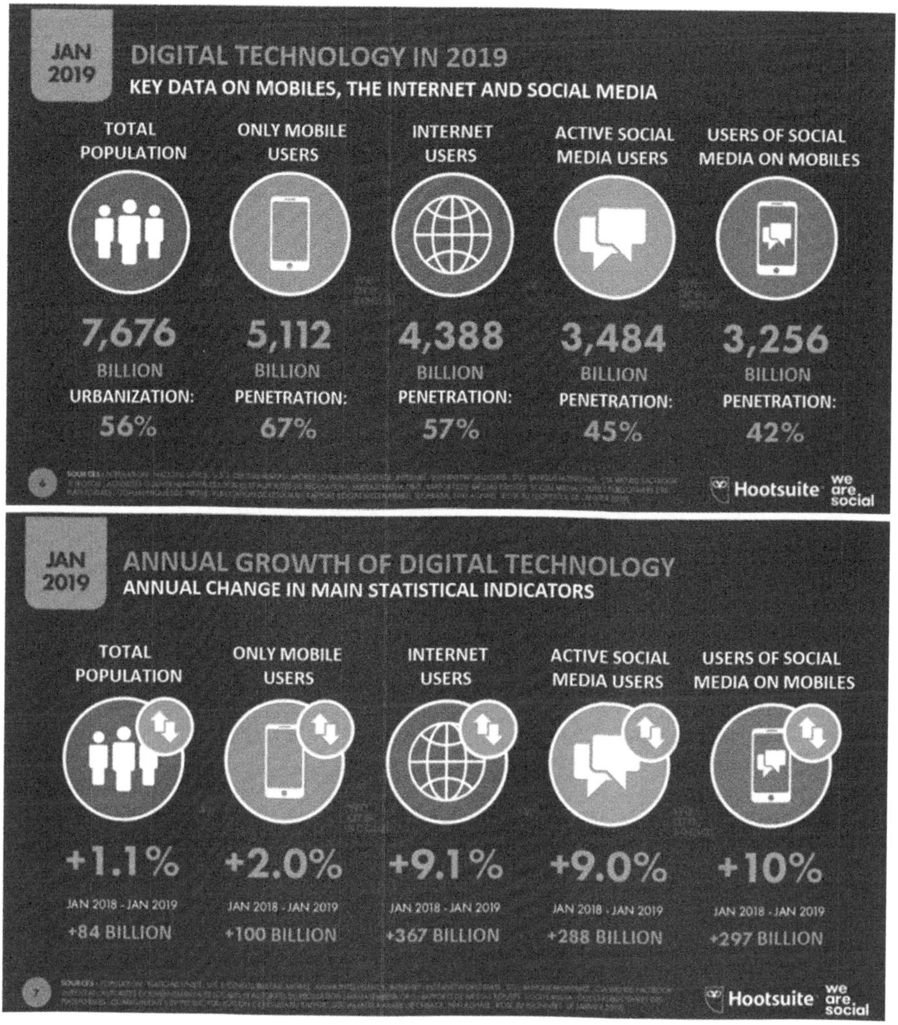

Figure 2.3. *Digital technology in the world and its growth (source: We Are Social/GlobalWebIndex). For a color version of this figure, see www.iste.co.uk/monino/control.zip*

In France, out of 65 million people, 83% are Internet users; 42% are registered on Facebook, i.e. 28 million members. More than 72 million phones are activated, and the French spend on average more than 4 hours a day surfing the Internet. French motorcyclists spend 58 minutes or more there; 68% of the population is registered on social networks. French people spend more than one and a half hours a day on social networks.

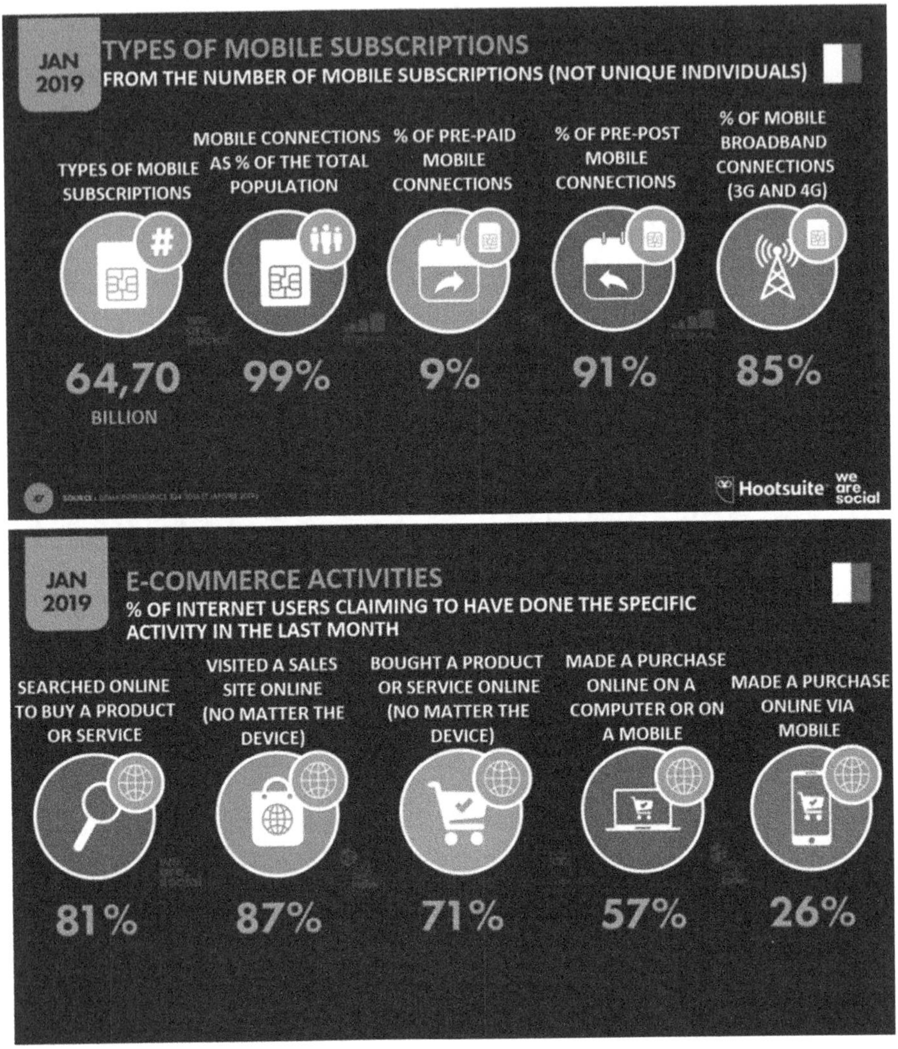

Figure 2.4. *Types of subscriptions and e-commerce activities in France (source: We Are Social/GlobalWebIndex). For a color version of this figure, see www.iste.co.uk/monino/control.zip*

Today[5], every individual creates, consumes and uses digital information. More than 3.4 million emails are sent worldwide every second, or 107,000 billion per year with 14,600 emails per person per year, but more than 70% are spam. Billions of contents are shared on social networks such as Facebook with more than 2.46 million shared contents every minute. The time spent on the Internet is more than 4.8 hours per day via a computer, and 2.1 hours via a mobile phone.

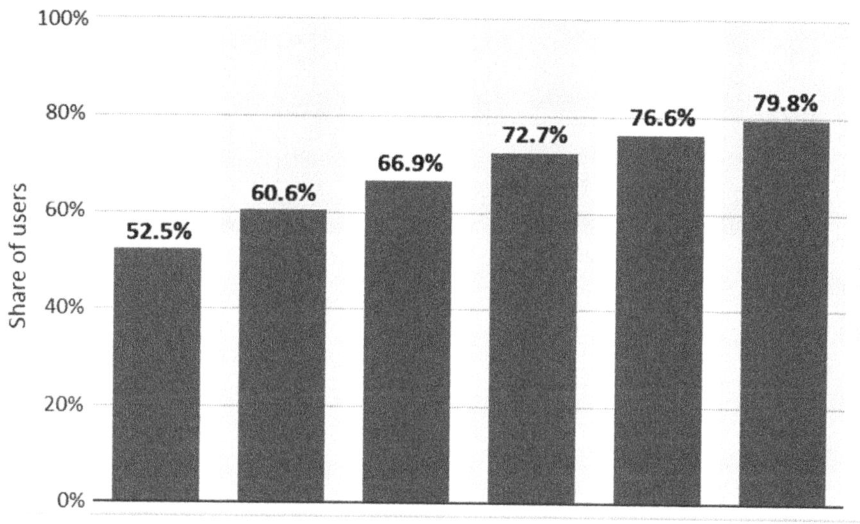

Figure 2.5. *Penetration rate of smartphones among mobile phone users in France from 2014 to 2019 (source: Statista). For a color version of this figure, see www.iste.co.uk/monino/control.zip*

This graph represents the penetration rate of smartphones among mobile phone users in France from 2014 to 2015, and presents a forecast of the evolution of this rate for the years 2016 to 2019. The source predicted that by 2019, the penetration rate of smartphones in France would reach around 80%. Since 2014, smartphones have supplanted "classic" mobile phones in France.

5 The international agency We Are Social and the social networking platform Hootsuite publish their Digital Report 2019, which delivers the latest trends in digital, social media, mobile and e-commerce around the world.

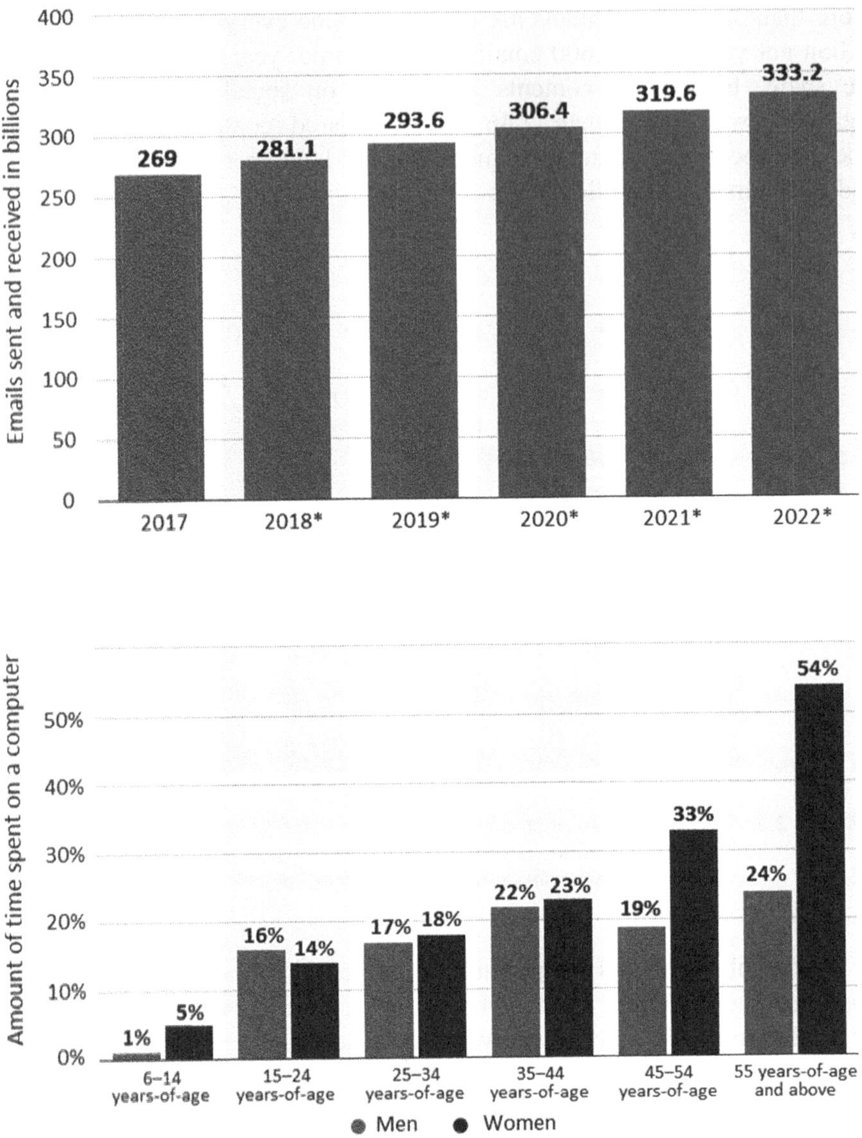

Figure 2.6. *Number of emails sent and received every day in the world from 2017 to 2022 (in billions). Breakdown of time spent on the computer by the French in 2015, by age and sex (source: Statista). For a color version of this figure, see www.iste.co.uk/monino/control.zip*

Figure 2.7. *Comparison by age of the number of hours spent on the Internet per week in France in 2019 (source: We Are Social/GlobalWebIndex). For a color version of this figure, see www.iste.co.uk/monino/control.zip*

The development of related technologies, such as the Internet, social networks, and cloud computing[6], has increased the amount of data available.

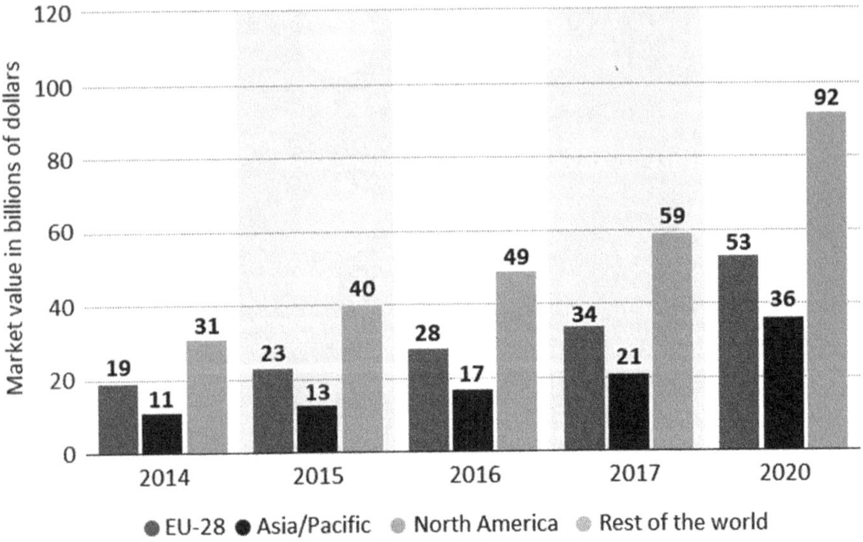

Figure 2.8. *Value of the Cloud computing market for companies worldwide from 2014 to 2020, by region (in billions of euros) (source: Statista). For a color version of this figure, see www.iste.co.uk/monino/control.zip*

2.1.1. *Figures on social networks*

The share of users of[7] social networks continues to increase, with 328 million more users between October 2018 and October 2019 (+9.6%). More than 97% of social network users access them from a mobile phone (3.660 billion out of 3.725 billion) and their number has increased by 15% year on year (+476 million). Finally, 123 million people connected to the Internet for the first time between early 2018 and early 2019 (+2.4%).

6 Digital factories.

7 Source: We Are Social/GlobalWebIndex.

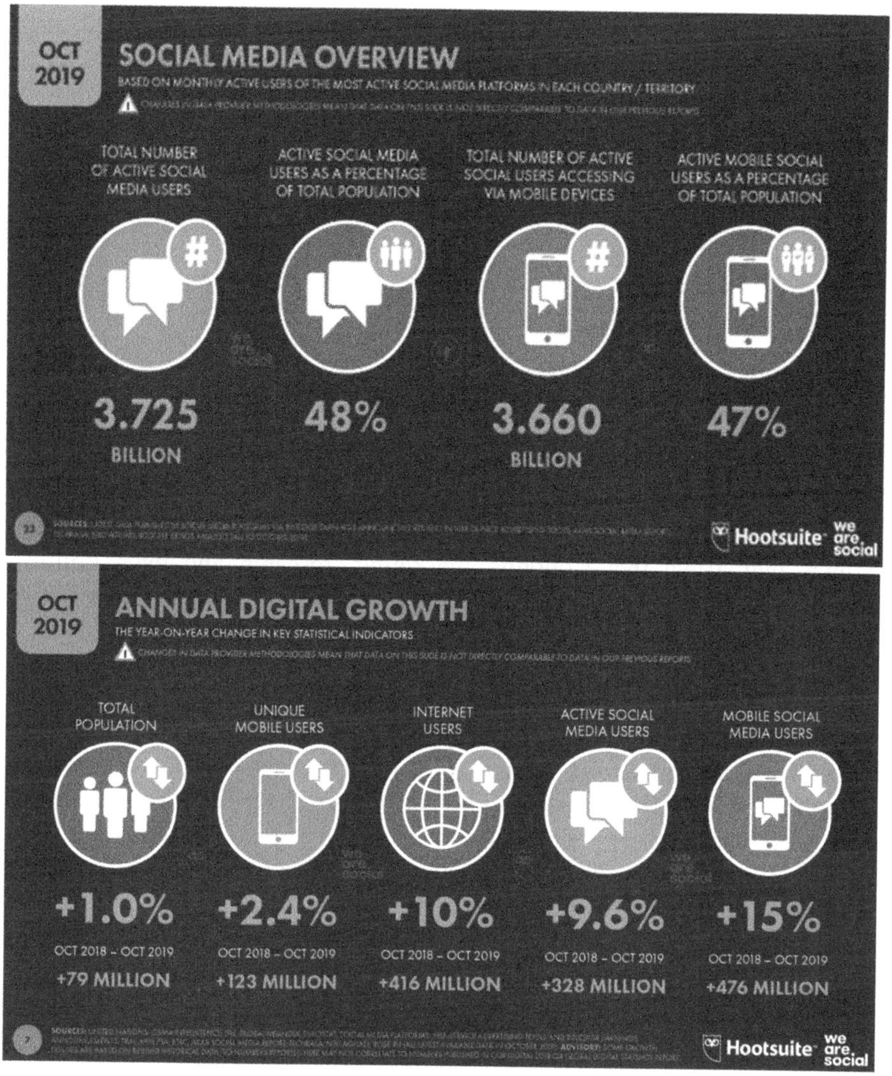

Figure 2.9. *Statistical social networks*[8] *(source: We Are Social/GlobalWebIndex). For a color version of this figure, see www.iste.co.uk/monino/control.zip*

8 Source: We Are Social/GlobalWebIndex.

Out of the 7.7 billion inhabitants, motorcyclists (5.1 billion, 67%) now outnumber Internet users (4.45 billion, 58%). There are 3.7 billion active users of social networks, representing 48% of the world's population, while 3.6 billion people connect to social media from a mobile phone (47%). Time spent on social networks: 3:32 in Latin America, 3:12 in the Middle East, 2:22 per day in Asia-Pacific, 1:57 in Europe (1:17 in France) and 1:56 in North America.

2.1.2. *Numbers: Big Data*

The development of these masses of data and their access represents what is called "Big Data". This immaterial data arrives continuously; their processing poses problems, particularly in knowledge extraction. Thus, new methods of automatic information extraction are being implemented: for example, "data mining[9]" or "text mining[10]". They underlie profound changes that affect the economy, marketing, research and even politics. The amount of data will increase sharply with the arrival on the market of connected objects that will gradually come into use.

Elements of our daily life are already connected: the car, the television and some household appliances. They are or will be equipped with a chip that collects and transmits data to their users via a computer, tablet or smartphone. The most important thing is that these items will also be able to trade with each other! This will allow us to remotely manipulate the equipment in our home, in our vehicle by connecting to them from our home and outside our home, using smartphones or any other equipment. This is called "the Internet of Things". The future "all-connected Internet" will result in the concrete fusion of the Internet, the Internet of Things and Big Data.

9 The term *data mining* refers to the analysis of data that consists of transforming data into useful information by establishing relationships between data.

10 Text mining is a set of techniques combining linguistics, semantics using statistical techniques and computer science. These techniques make it possible to recreate information from a corpus of texts (classifications, data analysis, trends, etc.).

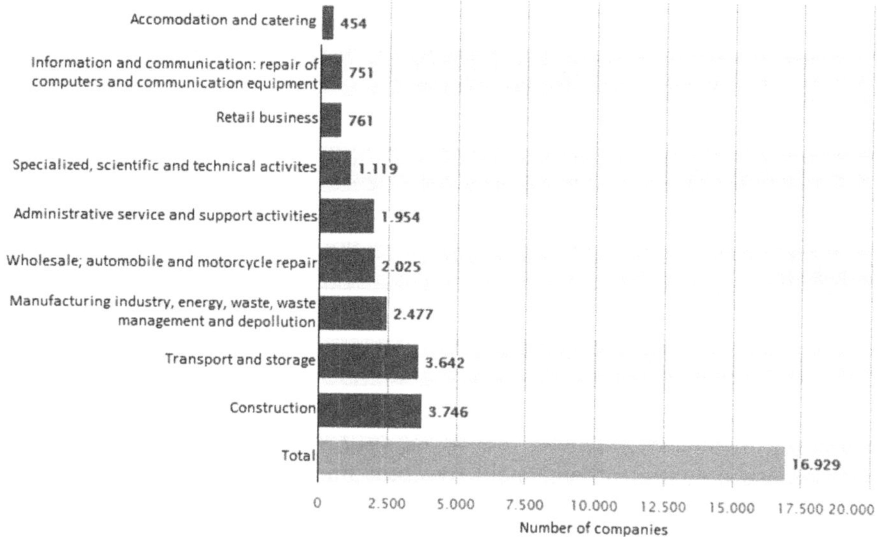

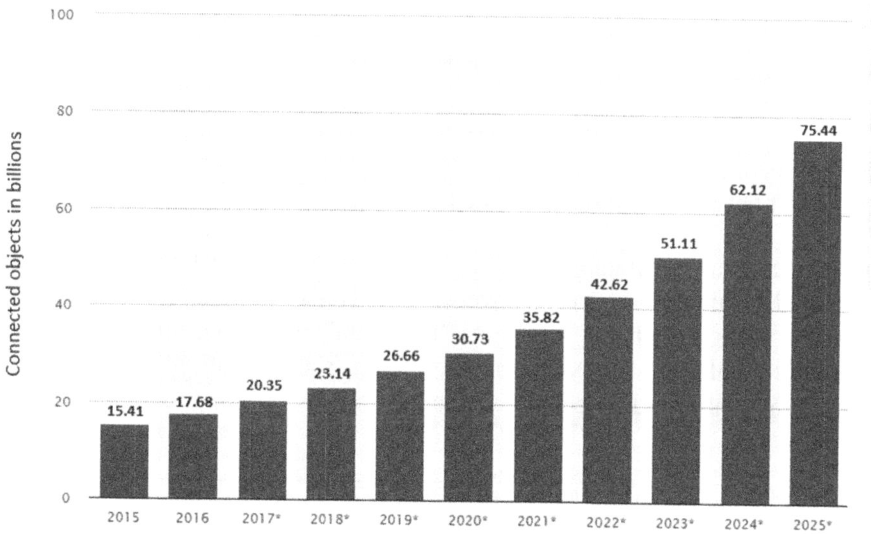

Figure 2.10. *Number of companies analyzing Big Data using geo-location data from handheld devices in 2017, by sector* Internet of Things (IoT): number of connected devices worldwide from 2015 to 2025 (in Billions) (source: Statista). For a color version of this figure, see www.iste.co.uk/monino/control.zip*

The American economist "Jeremy Rifkin" announced the advent of a new abundant society induced by the technological revolution, of which the new Internet of Things and 3D printing would be privileged drivers. New technologies are expected to change socio-economic relations, to the point of drying up the margins of capitalist companies. In the Internet world, the advent of the zero marginal cost society has already taken place. The dematerialization of information and the possibility of reproducing and distributing it at a marginal cost close to zero have profoundly changed the economic model of these industries.

Box 2.2. *"Zero Marginal Cost Economy"*

This phenomenon is now of interest to operational decision-makers (marketing managers, financial managers, etc.) when it comes to analyzing the immense potential of data held by companies in real-time. In order to meet the challenge of "Big Data", measures should be taken that include all the tools that allow data to be processed in a more restrictive way, as well as all the actors who analyze these data. This will only be possible through an awareness of the gains offered by "data enhancement". The data in the organized and reorganized bases, processed by statistical methods or econometric modeling, become knowledge.

2.1.3. *Key figures: the Internet of Things*

For a company, it is essential to have more and more data on the environment in which it operates or will operate. We will no longer work on classes of behaviors but on individual analysis.

It is easy to understand that this revolution is leading to the creation of so-called "start-up" companies whose aim is the automatic processing of data that make up what is known as "Big Data". This is certainly one of the components of what some call the "new industrial revolution". The Internet, digital and connected objects have opened up new horizons in a multitude of fields.

Access to the data enriches both quantitative and qualitative analyses. Customer contacts can be analyzed with data collected by a call center; this kind of product can also be offered in limited numbers, as E2S-Conseils does. It must be developed by exploring the content of emails, voice calls and mixing this information with website navigation. Or, study the messages exchanged on social networks (Facebook, Twitter, LinkedIn, etc.) to identify interesting new products or to know which products are most talked about

Box 2.3. *An example of data access*

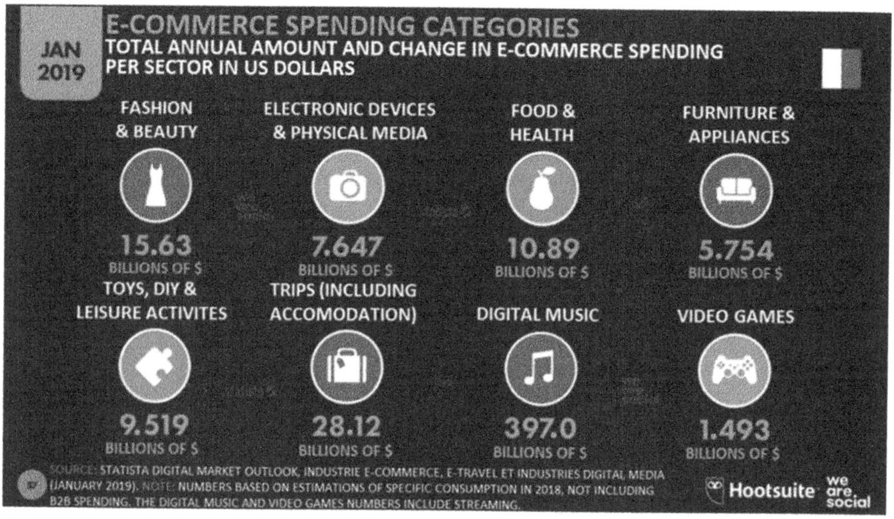

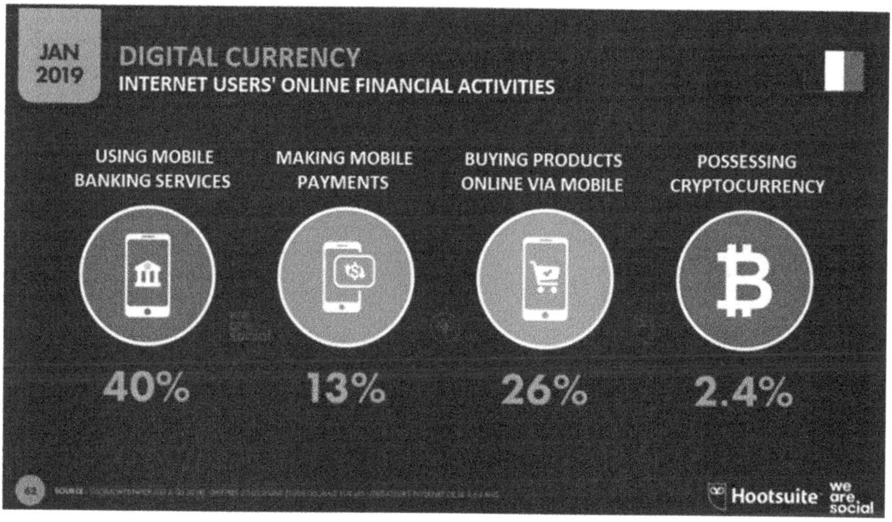

Figure 2.11. *Categories and digital currency (source: We Are Social/GlobalWebIndex). For a color version of this figure, see www.iste.co.uk/monino/control.zip*

In addition to the advent of ICT, the speed of production, dissemination and data processing capacity, another element has become important in recent years: time. This temporal element implies a notion of speed of information flow. This calls for a rethinking of corporate strategy, beyond the difficulties posed by the processing of large volumes of data. The value of a data item increases over time and depends on the multiple uses that are made of it.

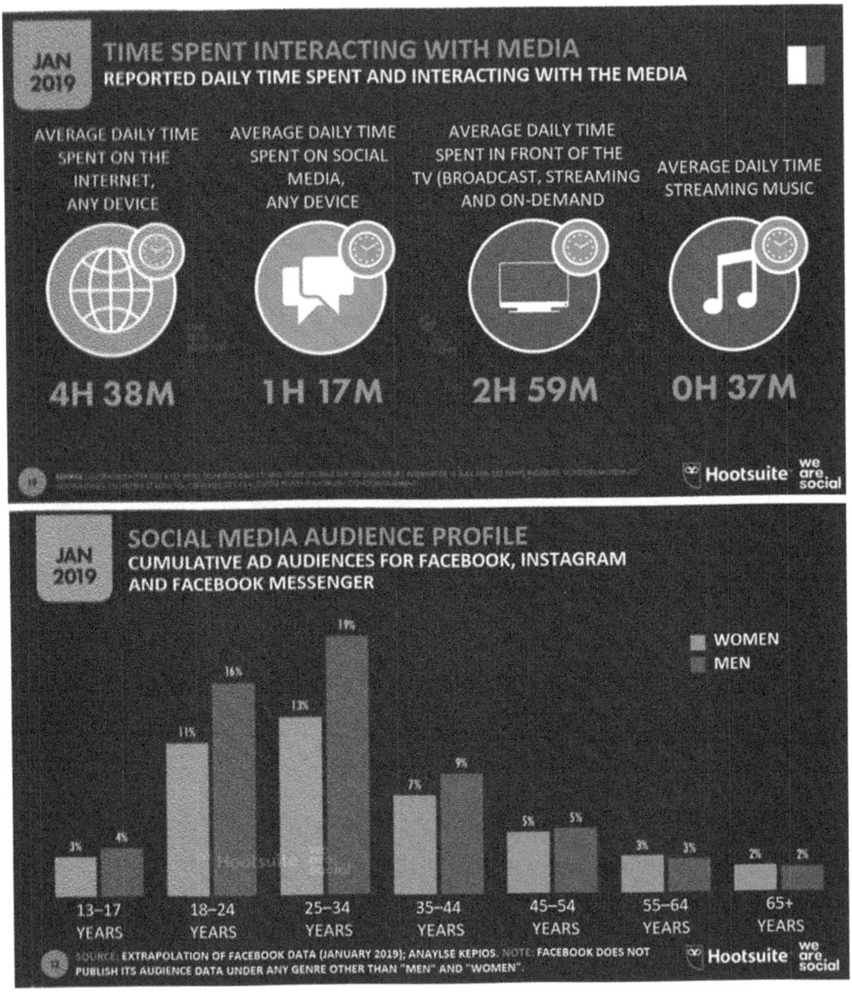

Figure 2.12. *Time spent and profile of the social media audience in France 2019 (source: We Are Social/GlobalWebIndex). For a color version of this figure, see www.iste.co.uk/monino/control.zip*

For data to reach its full value potential, it must be easily accessible by all interested parties without additional barriers and at economically bearable costs. If the data are open to users (Matouk 2014), other specific data processing companies may be created. This activity will meet the needs of users without the need for them to build models and equations themselves.

The process of collecting, processing and interpreting information is a practice that goes beyond the definition of ideas to the realization of those ideas to ensure a better production of knowledge that generates innovations. It is economic intelligence that enables each company to optimize its offer quantitatively and qualitatively and also to optimize its production technologies.

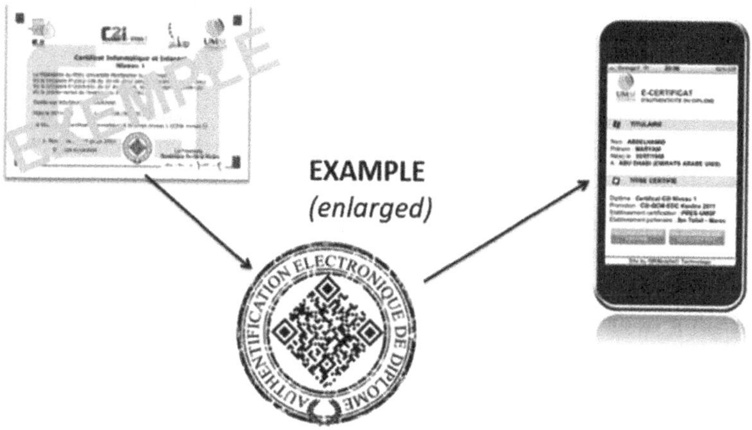

Figure 2.13. *Information processing. Security and mass processing of C2i certificates*[11] *using QRCode*[12]*. In recent years, multiple initiatives have been launched to certify the origin of diplomas. In this respect, it is interesting to recall that the University of Montpellier and in particular its TRIS Laboratory*[13] *have been pioneers in the fight against diploma falsification since 2011. Research work – https://www.contrefacon-riposte.info/technologies/4423-l-universite-de-montpellier-pionniere-dans-la-lutte-contre-la-contrefacon-des-diplomes. For a color version of this figure, see www.iste.co.uk/monino/control.zip*

11 C2i: Computer and Internet Certificate. Set up in 2004 by the French Ministry of National Education and Research.

12 The QR code is a type of two-dimensional barcode (or datamatrix code) consisting of black modules arranged in a square with a white background. The arrangement of these points defines the information contained in the code. QR (abbreviation for Quick Response) means that the content of the code can be decoded quickly after being read by a reader. (Wikipedia).

13 TRIS, *Traitement et Recherche sur l'Information et la Statistique* "Information and Statistical Processing and Research".

In this context, companies must have the capacity to absorb all available data, enabling them to assimilate and reproduce knowledge. This capacity presupposes the existence of specific skills that enable the use of this knowledge. The training of "data scientists" is therefore essential in order to be able to identify useful approaches to opening up or internal exploitation of data and to quantify the benefits in terms of innovation and competitiveness, since Big Data is only one element of a new set of tools and techniques called "data science".

The data scientist's mission is to extract knowledge from company data. He or she will be called upon to perform strategic functions within the Commission. To do so, he must master the necessary tools. He must also be more pedagogical and increase his mastery of data mining, because the volume of data requires an increase in the range of techniques to be mastered.

The massive flow of data or so-called "Big Data", which is generated by the Internet, social media, cloud-computing, etc., has developed very rapidly. This is leading to a rethinking of business strategy beyond the challenges of handling large volumes of data. The latter can be organized and transformed into information, which will then be transformed into knowledge by a cognitive or intellectual operation. But reaching the full potential of data depends on how it is presented; it must be used and reused in multiple ways without diminishing its value. This means making data available in the right form and at the right time to any interested party wishing to exploit and add value to it.

2.2. The power of data: a major challenge

Data sources such as smartphones, tablets, social networks and web services are multiplying. Once connected to the Internet, these intelligent objects can feed huge databases and communicate information to other objects and humans (Privat 2002). These data must be processed and valorized to become "intelligent" or "Smart". Intelligence, which can be trained through the application of analytical techniques, can provide essential insights that decision-makers will need for strategy development, operational performance, risk management and so on.

Data is an essential part of our digital society. These data can be organized and transformed into information which will then be transformed into knowledge by a cognitive or intellectual operation. This demonstrates today that the most important wealth is no longer material or physical. It is now based on the immaterial and driven by knowledge (Pinte 2006). Intangible heritage has overtaken tangible heritage (Monino and Lucato 2005, 2015). We produce them at all times and in all shapes. When data is contextualized and materialized, it makes it possible to formalize exchanges between individuals or between machines. This data can be paper or digital documents, images, videos, soundtracks, microfilms, informal discussions from person to person, etc.

Data has always had strategic value, but with the breadth of data available today and the capabilities to process it, it has become a new form of asset class. We are only at the beginning of a long journey where, with the right principles and guidelines, we should be able to collect, measure, and analyze more and more data to make better decisions, individually and collectively.

Today everyone produces data, sometimes consciously, sometimes unconsciously: humans, machines, connected objects and businesses. The data we produce and other data we accumulate are a permanent source of knowledge. Data is therefore a wealth of information and exploiting it provides an essential competitive advantage in an increasingly competitive environment. The accumulation of wisdom (intelligence and decision) is not truth. Wisdom appears when the fundamental principles on which knowledge is based are understood. For Eliot, wisdom is the ultimate phase of his poem. To introduce the concept of economic intelligence we have assimilated wisdom to decision-making (see figure below).

The advent of "Big Data" is changing traditional management practices and invites the organization to adopt new visions that allow it to have "strategic information" that can generate more value. In other words, the message must be conveyed that information per se is not power: it is using information that gives power, and the more information is given and exchanged, the more information is received (Martinet and Marti 2001).

This new immaterial manna that is "data" is produced in real-time, arriving in a continuous stream and coming from a multitude of sources that are generally heterogeneous. This accumulation of data of all kinds (audio, video, files, photos, etc.) generates new activities that aim to analyze these huge masses of data. It is then necessary to adapt and try to bring new approaches, new methods, new knowledge and new ways of working, which gives it new properties, and new issues since it is referenced in a disparate way. At the enterprise level, this mass of data is difficult to manage, and interpreting this large volume is mainly a challenge. This has an impact on companies that have to "manipulate" multitudes of data requiring a specific infrastructure for its creation, storage, processing, analysis and retrieval. The biggest challenge lies in "leveraging the data" available, in terms of quantity, diversity and speed of access. Today, data comes from all over the world: geolocation sensors, smartphones, social networks where we share files, videos, photos, purchase transactions on the Internet by customers, banking transactions through credit cards and so on. We must organize them to be able to use them.

As there is a lot of data, we must:

- collect them;
- store them;
- index them to make them accessible;
- edit them so that others can access them.

Companies are at the forefront of data repositories, and free access to this data gives them the ability to refine, interpret, and track trends as well as specificities.

Box 2.4. *How to organize and use data*

Data mining is a key issue for companies: knowing how to process data for better performance.

2.3. The Big Data revolution, "Mega Data"

The amount of data generated by people, Internet-connected devices and businesses is growing at an exponential rate. Financial organizations, businesses and health services generate large amounts of data during their interactions with providers, patients, customers and employees. Even outside of these interactions, a lot of data is being created through Internet search queries, social networks, GPS geolocation systems, stock market transactions and so on. The result is what is known as the "data revolution" or "Big Data" era.

The term "Big Data" is used to describe a universe of very large datasets that hold a variety of data types. This has given rise to a new generation of information technology to facilitate the fast processing speeds needed to analyze and extract value from datasets, using, of course, specialized hardware and software. The Big Data phenomenon is not only the explosion of the mass of data produced, made possible by the development of capacities for storing and disseminating information on all media. But it also refers to a second phenomenon, which is the new capacity to process data.

Generally speaking, the notion of Big Data allows us to refer to the current situation of wondering how to manage the mountains of data in a different way and how to make sense of this mass of data produced daily?

To do this, "data scientists" need to work together, at the scale of analysis and calculation of data warehouses, in order to gain clarity and precision. Data scientists must make data talk by using statistical techniques and specialized computer tools to organize, synthesize, and translate the information that companies need to facilitate decision-making.

2.3.1. *Understanding the world of Big Data*

In the literature the concept of Big Data is defined through the 4V or even 7V theory. The exponential speed at which the data comes to us as well as the multiplicity of sources that generate them in different formats (digital, text, image, etc.) are the characteristics of this phenomenon.

The "wave of computerization" that has swept over our societies is now reaching a form of maturity. Turning back to this flow, we cannot but be struck by the extent of the transformations that will have affected all sectors. Such a massive wave has developed very rapidly and has brought new uses. Information and communication technologies (ICTs) and the advent of the Internet have exploded the flow of information (Big Data or Mega Data). The world has become digital and technological advances have multiplied the number of data access circuits.

But what exactly is Big Data? This notion is an English-language concept that can be translated by the expression "Mega Data". This concept really took off with the dissemination of the three main reports written by the McKinsey Institute:

– Clouds, important data and intelligent assets: Ten technology business trends to watch (Bughin *et al.* 2010).

– Are you ready for the "Big Data" era (Brown *et al.* 2011)?

– The Big Data: The next frontier for innovation, competition and productivity (Manyika *et al.* 2011).

"Big Data" describes: "a set of data, data types and tools to respond quickly to the growing amount of data that companies process around the world." The amount of data collected, stored and processed by a diverse range of companies has increased exponentially. This has been helped in part by an explosion in the amount of data from web transactions, social media and sensors.

The growth of available data (Monino and Sedkaoui 2016), in quantity, diversity, speed and value, is explosive; these are the famous four "Vs": "Volume", "Variety", "Velocity" and "Value", by which English-speaking world define the term Big Data:

– Volume: the advent of the Internet, bringing with it a flood of social media updates, device sensor data and an explosion of e-commerce, means that every industry is overwhelmed by data that can be extremely valuable. All these new devices produce even more data and, in turn, add to the existing volume.

– Variety: with the rise of Internet and Wi-Fi networks, smartphones, connected objects, and social networks, more and more data in a variety of forms is being generated. These data come from different sources and are of varied nature (SMS, Tweets, social networks, messaging, etc.).

– Velocity: the speed with which this data is produced, made available and interpreted in real-time. The possibilities of real-time data processing are an area of particular interest, allowing companies to present achievements such as: the display of personalized ads on the web pages we visit, visualization and purchase history.

– Value: the goal of businesses is to capitalize on this deluge of data, including making sense of it.

Big Data issues are related to the volume of data, its variety, processing speed and value. Some researchers add three more Vs, namely:

– variability;

– veracity;

– viewing.

The first V refers to data whose meaning is constantly changing. The second indicates the result of the use of the data, because although there is a broad consensus on the potential value of Big Data, the data is virtually worthless if it is not accurate. This is especially true in programs that involve automated decision-making, or feeding data into an unattended machine learning algorithm. The last V, which is one of the challenges of Big Data, is in the way the results of the data processing (Information) are presented, which makes the conclusions clearer.

The expression Big Data represents a market in itself. Gilles Grapinet, former (2009) Deputy Chief Executive Officer of Atos emphasizes that: "with Big data, an organization's data becomes strategic, it is the discovery of a gigantic gold mine whose resources were previously unsuspected". This enormous amount of data is a valuable asset in our information society.

Big Data is therefore a broad discipline that is not limited to the technological dimension. Over the last few years, this concept has been arousing growing interest among players in information management systems. The concept of "4V" or even "7V", brings axes of reflection but does not give a clear definition of this phenomenon. The sum of these "Vs" opens up prospects for the creation of new products, better risk control and more refined customer segmentation. Actions to anticipate and reduce terminations and build customer loyalty are also possible.

Increasing data volumes, processing speed and diversity pose new challenges to companies and affect their decision-making processes. However, companies that generate, manage and analyze large datasets on a daily basis now use terms such as terabyte, petabyte, exabyte and zettabyte.

This phenomenon of Big Data has challenged traditional data processing methods and now represents an opportunity in the business world, especially for companies that know how to use it.

There are several methods by which a company uses its data assets to create value. The data can be used to develop a better understanding of customers and tailor products and services to their needs. The company can use the data to monitor the performance of key functions, identify the factors contributing to observed variances and highlight necessary corrective actions or new ways to optimize management systems.

Some companies combine data to predict customer behavior and therefore take action. Many other uses allow the company to better act on its environment.

The Big Data has therefore transformed all companies in all the sectors of activity, the way they operate and the way they act. We can also add dynamic and real-time analysis, thanks to the speed at which techniques and tools make it possible to obtain results. Analyses highlight changes in customer behavior and reveal new needs. They also allow the prediction of applications that do not yet exist in order to make strategic decisions.

Big Data allows companies to measure different aspects of daily life and correlate these measurements to discover relationships that the companies themselves could never have imagined. It gives the opportunity to look at a market composed of millions of customers and to see them not as a mass but as individuals with tastes and values. It provides companies with a statistical basis for identifying trends through data analysis tools.

The rise of Big Data reflects an awareness of the "power" of data and the need to better control the processes of collection, use, and sharing within companies. By facilitating more efficient decision-making, access to and processing of a large amount of information can help the company achieve a better strategic position. Data is therefore becoming the new strategic asset for companies, whatever their sector of activity.

2.3.1.1. *What changes in data analysis?*

Data analysis has always been a necessity for companies in order to have an accurate picture of their situation and to forecast their future economic activity.

Data analysis, before using the word "Big", refers to the discovery and communication of useful and effective models. It generally uses a variety of methods from different research backgrounds such as statistics, data mining and visual analysis and addresses a wide range of application tasks, including data summarization, classification and prediction, correlation, etc.

In the 1970s and 1980s, computers were used to process information, but they were very large and expensive. Only large groups were eligible to analyze the data. "Edgar F. Codd and Hubert Tardieu" were the first to work on data organization by designing database management systems (DBMS) and specifically relational databases. Today data processing and analysis is known as "Business Intelligence", thanks in particular to the increase in computing power of computers.

A fundamental requirement for successful data analysis is to have access to semantically rich data that links all the relevant information elements for objective analysis. However, the situation has changed with Big Data because the data come from several sources of different nature, in different forms (structured, unstructured). This leads us to say that new data processing tools and methods capable of combining thousands of datasets are needed.

In the world of Big Data, companies are eager to unlock the potential of data to generate value. They are also eager to find ways to process this data to make smarter decisions that will result in better service for their clients, improved efficiency for their processes and better results for their strategies.

"Big data refers to the volume, variety and speed of structured and unstructured data, transported through networks in transformation processes and storage devices until it is converted into useful knowledge for companies." (Gartner research firm)

The image below represents one of the visions of the mass of data and its increasing evolution through the different interconnected technologies. This volume of data is now available thanks to storage capacities that have increased while their cost has decreased significantly.

2.3.1.2. *Challenges around Big Data*

The extensive data collection is often created in real-time and its rapid processing enables employees and managers to gain access to previously inaccessible knowledge and to optimize their decision-making processes.

The data is then transformed into an action plan to be implemented, decisions to be taken, new markets to be penetrated:

– massive data storage and management;

– the management of unstructured data that often constitutes the bulk of the data, i.e. how to organize text, videos, images, etc.;

– the analysis of this massive data, both for reporting and advanced predictive modeling, as well as for deployment.

In common usage “Big Data” is not only used to refer to large datasets, it also gives way to the process of analyzing and extracting value from huge amounts of data. The term “Big Data” thus refers to the technologies, processes and techniques that enable an organization to create, manipulate and manage data on a large scale (Hopkins and Evelson 2011) and to extract new knowledge from it in order to turn it into economic value.

The vast amount of data collected, stored and disseminated by different processing technologies is shifting priorities and developing new analytical tools that align with changes in the way companies work to transform the business landscape. At the same time, new advanced analytical techniques are being used to interrogate datasets. The treatment of the latter will play a very important role in giving companies a competitive advantage.

This involves data acquisition, data cleansing and data preparation in different databases. Then we cross-reference this data to organize and load it with different tools to analyze it (to discover correlations). The data from these different sources are then combined into a single schema. That is, the data warehouses thus created are queried to present (visualize) the results in various forms.

In addition, there are now tools to process and assimilate this mass of data more quickly. However, mastering the technological dimension is fundamental because it allows us to know its limits and potentialities, and to make our interventions more relevant. With the multiplication of the quantities of data, companies are trying, through analysis tools, to understand how to extract value from the set of data collected.

A study by McAfee and Brynjolfsson (2012) showed that companies that have adopted advanced data analysis techniques achieve better productivity and profitability rates than their competitors. In fact, mastering data processing techniques is now becoming a real strategic and useful issue for the competitive differentiation of companies (Bughin and Chui 2010).

Box 2.5. *Study on mastery of data processing techniques*

Thanks to the new techniques and tools associated with Big Data, it becomes possible to work with large volumes of data. An advantage of this is that correlations with new data can be identified. The interpretation of this large volume of data is the main challenge of Big Data because the information generated can be the basis of new knowledge leading to opportunities for business development.

Today's technologies offer a structured and static vision of data. Performance and storage limitations reduce the possibilities of analysis to explicit datasets. Most solutions use a "Storage-Processing" duo. It should be noted that the increase in the volume of data has been accompanied by a decrease in the price of data storage.

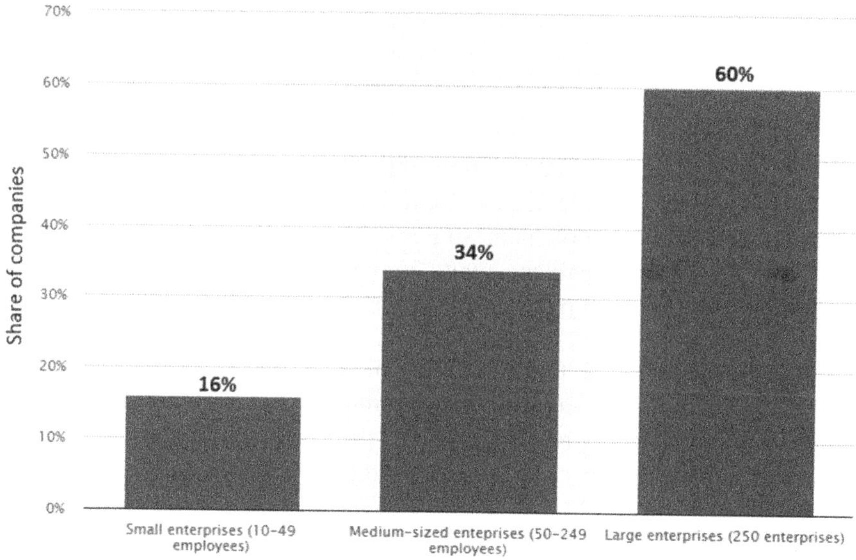

Figure 2.14. *Percentage of companies purchasing Cloud computing services used on the Internet in France in 2018, by company size (source: Statista). For a color version of this figure, see www.iste.co.uk/monino/control.zip*

Currently, one of the innovations, making it possible to store and share large amounts of data, is "Cloud Computing", or "the Internet in the Cloud". The Cloud enables access to shared IT resources via an on-demand or self-service telecommunications network. The Cloud transforms storage infrastructures and computing power into services, through companies that own servers and lease their capacity. This approach makes it possible to mutualize needs and give users great flexibility in terms of data storage and processing speed.

2.3.1.3. *Making data warehouses intelligent*

But how can we obtain or increase these gains based on the masses of data? How can companies gather and combine data from disparate sources to achieve this vision? What role can data analysis play in what is essentially an IT challenge? What changes may be required to make data analysis a more practical discipline? All of these issues represent the biggest challenge or the difficult part of this change, making Big Data a "Big challenge".

> The McKinsey Global Institute report (July 2013) estimates that the implementation of Big Data practices could generate $30 million in additional revenue in the retail sector through productivity gains, improved transparency and more sophisticated targeting of marketing, advertising and offers.

Box 2.6. *The McKinsey Global Institute Report on Big Data and Smart Data*

The challenge of Big Data is now the intelligent management of data warehouses, in order to identify and extract relevant information allowing companies or users to implement strategies that really meet the identified needs. Intelligent data enables you to move from raw data, structured or unstructured from internal or external sources, to strategic information.

Data quality refers to the suitability of data for its intended uses, in operating procedures, processes, decision-making and planning. Similarly, data quality also plays a role in product life cycle analysis. It is of great importance and the data is an important source of value for companies.

Data quality is important for monitoring and evaluating progress towards the achievement of objectives. It is all the more so when it relates to accurate and reliable information collected through a company data management system.

Accurate information is needed to:

– demonstrate accountability and good governance;

– provide decision-makers with the information necessary for planning, resource allocation, strategy development, etc.;

– monitor progress towards the achievement of established goals and objectives.

Indeed, companies should not rely on the size of their data; data is only useful if it is used intelligently. So, it does not matter how big or small because internal data must be combined with external data in order for a company to get the most out of its data assets. These include excellent analytical skills, an ability to understand and manipulate large datasets, and an ability to interpret and apply results.

The challenge is to refer to the use of data, rather than quantity. This could become the most cost-effective way to ensure that the value of the data is contained within the available masses. The change from "Big Data" to "smart data" symbolizes the consideration of meaning in data processing. "Data scientists" appear at this level, who are competent in computer science, mathematics and statistics, with a good knowledge of the business world. They must be able to analyze a phenomenon from every possible angle in order to take advantage of the company's data assets.

Thus, under the term "variety", there are many problems. First of all, data, especially in an industrial environment, can be presented in various forms such as texts, functions, curves, images and graphs, or even combinations of the whole. However, these data are observed with great variability, which often masks the nature of the phenomenon being studied. It is therefore important to reveal the structure contained in these observations in order to extract the right information from them.

The processing of large masses of data, associated with the "new" profession of data scientist, occupies an important place in companies that have placed themselves in the context of the deluge of data requiring specific processing. The development of computing and computing resources allows the storage of databases, processing, and analysis of very large datasets.

More recently, the improvement of software and its interfaces has offered users (whether statisticians or not) very simple possibilities of implementation of these methods. This evolution, as well as the popularization of new algorithmic techniques (neural networks, machine learning, bootstrap, etc.) and graphical tools, leads to the development and marketing of software integrating a subset of statistical and algorithmic methods used under the terminology of "data mining".

Thus, data mining is presented as the search for relevant information for decision support and forecasting. It implements statistical and machine learning techniques, taking into account the specificity of large to very large data dimensions. But in the world of Big Data, the main objective is still the one related to a much more important "V", that of valuation.

The consensus today is to define the data scientist at the intersection of three areas of expertise:

– computers;

– statistics;

– mathematics.

Box 2.7. *The data scientist*

Let us discuss "Business Knowledge" (Abitboul *et al.* 2013). In relation to these new data structures, the statistician who has become a "data scientist", revisits his classics to focus on tools and methods leading to useful executions that are compatible with the new information systems.

The increasing production of data, generated by the development of information and communication technologies (ICTs), requires greater openness and useful sharing that is becoming a powerful force for change in the economic and social world. "Open data", another term that is part of the data wave, has spread around the world despite its novelty, thanks to its ability to generate both economic and social value. This phenomenon has attracted much attention in recent years and one of the central principles of the open data Initiative is to make public data accessible to promote innovation.

Moreover, for many functions, professional life requires reading and understanding a wide variety of data: market research, dashboards, financial statistics, forecasts, statistical tests, indicators, etc. For this purpose, various statistical techniques are used in the professional world to solve a wide range of problems, from customer relationship management to preventive maintenance, from fraud detection to website optimization. These techniques are known today as "data mining".

2.3.2. *Open data: a new challenge*

The world of data evolves rapidly on an axis of increasing complexity that can be expressed in volume, variety, value and so on. We then speak of data warehouses or Mega Data (Big Data). Data are essential assets for value creation, and a factor that promotes and stimulates innovation, growth, and development. With the digital revolution, data is also taking on a central role in the economy. But reaching the full potential of data depends on how it is presented; it must be used and reused in multiple ways without diminishing its value.

This means making data available in the right form and at the right time to any interested party wishing to exploit and add value to it. Open data adds a new dimension to data warehouse analysis and leads to new forms of innovation. Opening up and sharing data means organizing the online availability of essential data that will enrich the analyses of many decision-makers; it means saving working time, or better-informed decision-making in the various sectors. This means creating major reference systems shared by all the players and encouraging the development of numerous high value-added services.

2.3.2.1. *Why open data?*

Open data is digital data of public or private origin. It is produced by a local authority or a public service (possibly delegated). It is disseminated in a structured manner according to a methodology and an open license guaranteeing its free access and reuse by all, without technical, legal, or financial restrictions. Open data encompasses a number of sources and types of data:

– public data or information from the public sector. They include all data collected by public bodies at all levels;

– data from scientific research, particularly public research;

– private sector data that, with appropriate incentives and protection of information, can be released.

Open data is therefore the provision of this data in such a way that it is accessible, usable, and reusable by all interested parties (companies, researchers, etc.).

The philosophy carried by this movement is, at its origin, deeply civic. Thus, for citizens, free access to data contributes to the strengthening of democracy. This event should serve to enrich the democratic debate, energize public life and contribute to reinventing public services. Opening up data is a political process with a message built around transparency, innovation and the construction of public action.

It should be noted that open data has been in use since the 1970s, referring to the processing and modeling of statistical data that made data open and allowed their communication and transfer (Buckley, 1998). The history of the development of open data is summarized in Figure 2.15.

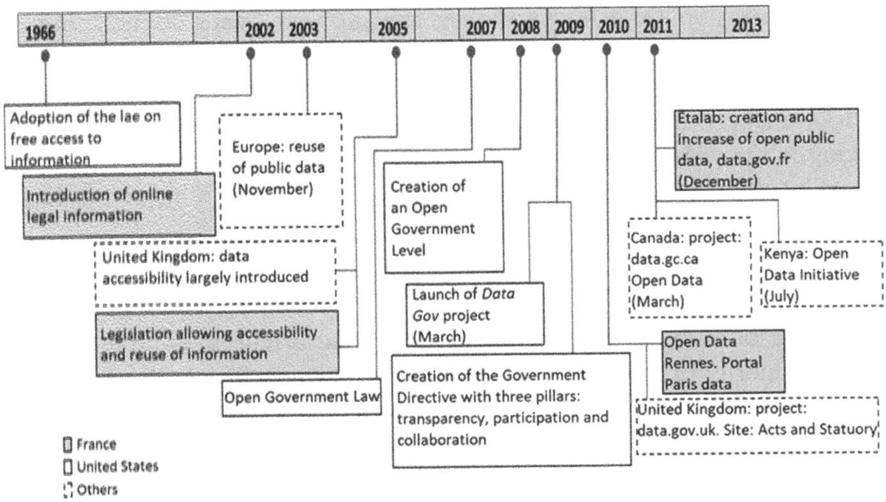

Figure 2.15. *History of open data (source: Monino and Sedkaoui). For a color version of this figure, see www.iste.co.uk/monino/control.zip*

After several years, France's strategy of opening up and sharing public data has taken shape with the launch of the data.gouv.fr portal in 2011. Since January 27, 2011, Paris has been one of the first cities to consider the issue with the launch of the "Paris Data" platform, which makes it possible to disseminate public and cartographic data. The city of Rennes has also played an important role by being the first to really work on the idea of making open data a vector for improving public services: by developing services and uses and by enhancing public data to improve citizenship through transparency of public data.

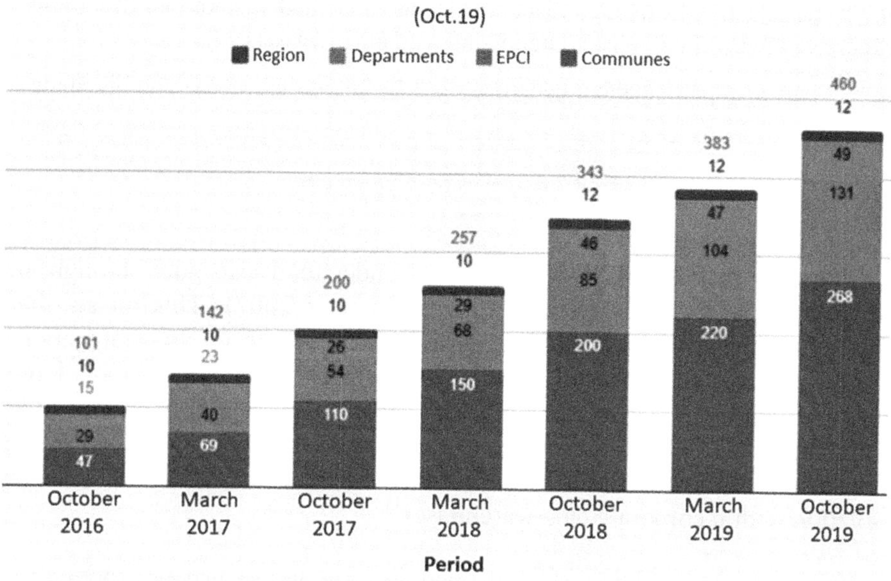

Figure 2.16. *History of communities that have opened their data*[14] *(source: Coudercy and Delobelle 2014). For a color version of this figure, see www.iste.co.uk/monino/control.zip*

Open data thus means that the public sector is relinquishing its role as a custodian of data and replacing it with that of a provider of data and information, leading to a realignment of power dynamics between the public and private sectors. This openness will enable a relationship to be built around data, between communities, businesses and citizens by encouraging

14 On the occasion of the presentation of the 2019 edition of the open data observatory, Laure Lucchesi, head of Etalab, reaffirmed the State's full commitment to the open data dynamic. This support is all the more expected as the need for support for small communities remains as great as ever, with barely 10% of communities having opened their data. https://www.banquedesterritoires.fr/la-dinum-reaffirme-son-soutien-lopen-data-local.

transparency, the development of new uses, and the participation of individuals in the enrichment of data.

The definition of open data gives precise details that mean:

– availability and access: data must be available in their entirety and also available in a convenient form and modifiable;

– reuse and redistribution: data must be provided under conditions that allow reuse and redistribution including mixing with other datasets;

– universal participation: everyone must be able to use, reuse and distribute the data.

Box 2.8. *Definition of open data*

The data that is linked, analyzed and understood also goes through an incremental cycle from creation to use (Mercier 2007). This means that the value of the data is not intrinsic but derives from its aggregation, cross-checking, analysis and reuse. The Internet and intelligent tools have thus updated a "Universe" of data that plays an important role. This importance was built around the exchange and fluidity on the Web. Data often has more value as a tool for exchange and influence than in isolation within corporate data warehouses[15].

The emergence of the reuse of open data (public or private), driven by new uses, has thus revealed a value chain based on these data, which when made available in a raw and reusable way, allows the creation of new services. Companies such as "Amazon" or "Netflix" have, as previously mentioned, taken advantage of the mass of users using their sites to extract patterns of consumer preferences, which allow them to make highly personalized recommendations according to the tastes of their customers.

Open data, both public and private, means more and better access to data. Quality criteria can be presented around: completeness, consistency, and accuracy. In addition, there are the dimensions of time and territory and interoperability through open formats. These principles are based on

15 Ed Parsons – Geospatial Technologist for Europe, Middle East, and Africa at Google, at the seminar "Innovations and development of the geographic information sector" organized on October 5, 2009 at the Grande Arche de la Défense by the Conseil National de l'Information Géographique (CNIG), AFIGEO and the Institut des Sciences et des Techniques de l'Équipement pour le Développement (ISTED).

important values: accessibility, autonomy, sharing and freedom (Mercier 2014).

Open data must meet three criteria:

– technical: the raw data must be automatically exploitable and made available in formats that are as open as possible and non-proprietary;

– legal: licenses must clarify the rights and obligations of data holders and re-users, and be as open as possible;

– economical: little or no charges (which may act as a brake on re-use), maximum marginal cost pricing, and so on.

Box 2.9. *Three criteria for open data*

Data is increasingly emerging as one of the structuring elements of today's digital world. In order for this data to generate knowledge, it must be accessible to all and made available online. They must be available to all users who can combine and mix them to create value. To take full advantage of open data, it is necessary to put data in a context that creates new knowledge and offers powerful services and applications.

2.3.2.2. *Making the data reusable*

Today, at the heart of the transparency issues linked to the context of the modernization of public action, it is more particularly the opening up of "open data", alongside "Big Data", of course, which is in the process of evolution. The more open the data is, the more it can be used, reused, redirected and combined with other data for the benefit of all. As long as our economy and society become more knowledge-based, data are essential assets for value creation and a factor in fostering and stimulating innovation, growth and development.

Transparency, participation and collaboration are the main issues at stake in the integration of the different economic actors in the open data paradigm.

The process of reusing open data involves several types of actors.

– Public actors: administrations, local authorities and public administrative establishments produce and receive a considerable amount of data.

– Companies: by their nature, companies use data in different internal processes to support their strategy. Companies thus produce data (reports, etc.) that they can combine with other data from the external environment (competitors, customers, suppliers, Web, etc.).

– Scientific research: researchers are by nature data collectors and even producers of data. The world of science is also governed by the principle of publication: recognition of a scientist is based on the results he publishes and submits to the scientific community.

– Individuals: each individual can retrieve data and enrich it with intelligent tools (smartphones, tablets, GPS, etc.).

Box 2.10. *Producers and re-users, types of actors*

The reuse of data is, of course, aimed at the most established players in the economy, especially large companies in the information sectors, who can use public data to improve their production processes. Because it represents a strong potential for innovation, public data is a particularly important ingredient in the development of start-ups and young innovative companies in the digital economy.

While companies are emerging to exploit the data made available by government, businesses and researchers or even citizens, so far few companies have embraced the open data movement themselves by publishing some of their data in open access. However, they are producers of data through their activity, and often consumers of external data that enable them to analyze the economy or to exchange with their partners. For them, the openness of data is more dependent on opportunities in terms of business or innovation than on transparency issues, although it also allows them to improve their image.

However, data openness goes beyond the realm of public data and scientific research data. A recent study by McKinsey Global Institute suggests that open data, and public data, have the potential to unlock about \$3.2 trillion in value added per year for seven different areas of the global economy. With the high-speed development of the mobile web, technologies allow for the processing of large amounts of data.

The opening up of data can generate several advantages for companies, namely:

– accelerating product development;

– stimulating innovation by creating a dynamic ecosystem of partners while improving customer satisfaction;

– understanding market changes.

Box 2.11. *The benefits of opening up data*

For example, in the field of health, the *Caisse Nationale d'Assurance Maladie* (CNAM) has a gigantic stock of data with the billions of sickness sheets it processes and stores the figures. These data are not yet available. The CNAM is resisting the request to open it under the pretext of medical secrecy. In fact, it can be perfectly preserved, using existing patient identity coding. The anonymity of doctors can also be preserved, although it would be very useful to expose doctors who have abusive practices, not so much to punish them as to lead them to change their practice, which is obviously not unrelated to balancing the health insurance budget.

If these data are finally open to honest users (Matouk, 2014), i.e. excluding ill-intentioned "insurers", and even more so "gutter journalists", health policies could be greatly enriched and compulsory health insurance budgets much better kept.

But the "mine" of data is also increasingly fed by another category of very specific data, which is often collected without our knowledge by multiple organizations: personal data. Our civilities, ages, income brackets, tastes, more or less frequent purchases, motives for these purchases and so on constitute a growing part of the "Big Data", but a mostly hidden part, which some users alone reserve the use of for their greater profit.

The data from the Big Data comes from:

– newspapers, social networks, e-commerce, indexing, storage of documents, photos, videos;

– genomic science, astronomy;

– business data;

– personal data, bypassing the GDPR;

– public data through open data[16].

Information management has become a crucial factor for business success. Effectively processing available data (public, private) into information is the key. With a combination of methods and technologies, companies can achieve and benefit from a competitive advantage.

It is true that information (which is originally data) becomes a strategic issue and a power for the party holding it. But it must be considered as an element of performance in the company because it represents a product of values generated by sharing and exchange. Open data tends to support this conception of power based on possession, to move towards a conception based on sharing.

The more open the data is, the more it can be used, reused, redirected and combined with other data for the benefit of all. Different business models are possible to make the best use of open data. Because exploited data generates billions of dollars according to the various reports, open data has become a planetary movement. It is a movement that claims free access and exploitation of data by all and for all.

Opening up and sharing data means organizing the online availability of essential data, which will enrich the analyses of many decision-makers, allow for savings in working time or enable better informed decision-making in the various sectors. The opening up of data must be seen as an opportunity and a long-term development for the various actors (administrations, companies, researchers, citizens).

The opening up of data represents a major source of information for all players, and for this reason it must be considered as a tool for rationalizing public action and company activities. Open data also represents the means to renew the methods of implementing public action and to develop links between (and with) administrations, businesses and citizens, because it is not a question of opening up public data simply for the sake of transparency.

16 The opening of data that intersects with the phenomenon of Big Data, which represents massive amounts of data, is a matter of course, in order to make them valuable in terms of utility. Making data open means putting it online, which calls for a new concept "Linked data" or "the web of data". The latter represents a growing movement for companies as they need to make their data available in a machine-readable format. This allows users to create and combine datasets and make their own interpretations. Web of data is about exposing this structured data on the web and linking it together, increasing its visibility and reuse.

2.3.2.3. *Open data and Big Data*

The Web is the largest and most dynamic data repository in the world and is the ideal resource for making the most of data and turning it into information. The data on the Web is very diverse and voluminous (content and format). The most important attribute of large volumes of data is that they allow knowledge to be put to good use and create considerable value. Combined with advanced analytical methods, they could shed new light on the behavior of several phenomena.

There are two ways to turn data into valuable input for a business:

– the transformation of data into information is a real stage of data value, which is exploited in order to have useful information and to carry out corporate strategies;

– the transformation of data into products or processes brings value to a company.

The point of the data revolution is to enable the development of a competitive advantage. Big Data is today a real ecosystem because it irrigates all sectors. Large companies and start-ups need to work together. Exchanges between experts, researchers, start-ups and large companies will develop a synergy leading to new knowledge and new projects.

The State thus has a crucial role, through the provision of public data and the adaptation of the regulatory framework, to facilitate access to and use of data in order to create value. Nevertheless, when opening up these raw data, public administrations must make them understandable to all.

The openness and easy access to data for all parties offer several advantages for both holders and users, namely:

– rediscovering other uses, reuse, linking, cross-referencing, and combining data with other forms of data from other sources;

– fostering innovation by providing easy access to data;

– gaining "feedback" by opening up the data, as a means of complementing internal analyses with ideas from external sources;

– increasing transparency which will lead to increased reliability of scientific results and increased credibility of companies;

– benefitting from the effects of the Web by combining several elements because each new data opening adds value to the whole ecosystem.

The fundamental principle of the open data philosophy is free access to data, which will facilitate use and reuse. Re-users will be able to cross reference them in order to produce new information that will better meet the expectations of companies.

An open data record must allow us to:

– create new products or services;

– access aggregated and updated information;

– enhance the company's image;

– exploit and develop new information from these data;

– analyze in order to make sense of the data.

Data accumulation is not information, it is numbers, words, events existing outside a conceptual frame of reference. In the absence of context, the individual data do not have much meaning. The data is raw information.

The question for a company is no longer whether to go ahead, but to take advantage of all the available data (structured or not) and to know how to adapt the different key success factors to its environment. These databases will not be limited to internal data (customer data) but will include all external data (communities, universities, etc.) revolving around the company to increase the value created by a data.

Optimize your production process, refine your knowledge of customers, control your reputation, rationalize your procurement costs, stimulate research, etc. The possibilities seem endless in the world of the data revolution. We must therefore look more for innovation in new economic models, in order to seek out economic performance and capture value.

The objective of an open data policy is to encourage creativity, stimulate innovation and promote the reuse of both public and private data, based on collective intelligence and the willingness of companies and researchers to create new knowledge capable of generating innovations.

The nature of the data allows it to be used, reused, linked, and recombined in multiple ways without diminishing its value. Open data supports the emergence and realization of large data potentials, and its main effect lies in the variety of sources. Although the volume of data is increasing, as well as the speed of processing and exploitation, open data is contributing to the search for new ways to add value to existing data.

In 2011, France took inspiration from the American portal data.gov launched in 2009 to create its own public data dissemination system, managed by Etalab to accelerate the movement. But the available data has a strong economic value for innovative entrepreneurs already. However, administrations are not the only ones to have embarked on open data. Etalab's mission is to support the opening and sharing of public data, in order to facilitate its reuse by companies, citizens, and other interested parties.

Etalab has launched a series of competitions in 2012 and 2013 to create innovative projects and services to encourage the reuse of public data. This is the "Data connections" initiative, to distinguish the best applications, services, or interactive data visualizations, notably reusing public data: six start-ups have been awarded prizes. Etalab also contributes to highlighting the best reuse of data, in particular by promoting it within the State.

In France the opening and sharing of public data is free, freely reusable and in open formats. This is already a resource used by hundreds of start-ups developing new value-added services. Even if public support for innovation through Big Data is present in France thanks to the opening up of public data and financial aid to start-ups and SMEs, few strategies are implemented to make the most of this data[17].

Once the identification of useful data has been carried out and the criteria for their quality have been analyzed, the data can be organized and transformed into information. The data generate information, and as such they admit a rather strict methodology:

– define precisely, what it is necessary to remember;

– provide for the archiving of raw and processed data;

– keep as much instrumental information as possible to keep the memory of the individual quality of the processed data;

17 See https://www.data.gouv.fr/fr/datasets/statistiques-open-data-bpce/.

– organize *a priori* and not a posteriori, the computerized constitution of the catalogue of observations and facilitate the future use of the data, by introducing additional fields to select and identify the observations;

– establish and maintain synthetic grids to quickly retrieve data.

Box 2.12. *Proposed approach*

The data revolution will necessarily involve Big Data and open data. It is therefore essential to address these issues and their definition. Both concepts can transform the world of business, government and society. Big Data gives us the power to understand, analyze and ultimately change the world we live in; open data wants this power to be shared and the world we live in to become more democratic.

Big Data concerns methods of collecting and processing a large volume of data in real-time; it is of great richness. Open data is a way of accessing information. The opening of data is a fundamental trend in data access: unlike most information in digital form, it is made available to the public or to professionals, without compensation. Opening up public data makes it possible to respond to a demand for innovation of an economic, scientific, environmental or social nature.

However, the latest reflections on the opening of data indicate two needs on the part of the re-users:

– more raw data updated in real-time, and processable;

– more context documents to understand how and why a datum has been constructed.

The opening up of data represents a major source of information for all players, and for this reason it must be considered as a tool for rationalizing public action and company activities. It is also a means of renewing the methods of implementing public action and developing links between (and with) administrations, businesses and citizens: it is not a question of opening up public data for the simple purpose of transparency.

Big Data and open data are closely related but they are not identical. Open data brings a perspective that can make Big Data more useful and democratic. Big Data is defined by size, which is the term used to describe very large datasets. But these judgments are subjective and technology-dependent: the amounts of data available today may not seem so great in a few years, when data analysis and computer technology improve.

Generating value at the different stages of the data value chain will be at the center of the intersection of openness and large amounts of data. Big Data and open data are establishing themselves as a source of information that is still under-exploited.

2.3.2.4. *Valuation and reuse of data*

The Web is entering a new phase of its existence, one of the characteristics of which is the quantitative and qualitative leap in the data placed on it. Data sources are diversifying: public authorities, companies, individuals, and tomorrow it will be around the objects of publishing, sharing, and circulating more and more data across the Web. Over the past two years, voices have been raised to promote a freer flow of such data.

Speed and ease of access to data is therefore crucial in a world where the amount of information available is constantly increasing. We can distinguish three main categories of Web-based data depending on the producer of the raw data, from which we propose the following definitions:

– raw data produced by public entities, i.e. data from the government, local authorities, statistical institutes, etc.;

– raw data produced by private or public enterprises that can be sold or made available free of charge;

– raw data produced by individuals, such as personal data. However, since May 25, 2018, the GDPR must protect individuals.

The relationship between different types of data (statistical, scientific, administrative, geographic, Web, etc.) allows the creation of ecosystems integrating a very large number of data from a multitude of different datasets. The reuse of open data plays a major role in terms of its value.

With the digital revolution and the advent of the Internet facilitating the production, dissemination and processing of mass data, the opening up and sharing of public data has become a powerful lever for:

– strengthening individual confidence through greater transparency in public action;

– enabling new forms of co-production with society and support social innovation;

– improving administrative operations;

– improving the effectiveness of public action by developing new organizational methods and new work processes;

– supporting economic dynamism by creating new resources for innovation and growth.

The process of understanding information is often seen as a bottom-up progression from data, to information, to knowledge and finally to wisdom. In the same way, open data are more important, they are more functional through their collection, structure and dissemination. In this way, data assembled in a relevant way reconstruct an informational world and structure a world of digital data.

Opening up the data not only leads to an improvement in accountability and trust, but also for entrepreneurs to innovate through the development of applications, most often in the form of "mashups[18]", i.e. digital products created from pre-existing elements. After identifying, collecting and sorting the data, a processing phase takes place in order to provide cleaned data ready to be used for analytical purposes. This processing phase is important because it allows the maximum amount of information to be extracted from the data at stake.

The growth of available data in quantity and value is the modern answer to the ever-increasing need of people and organizations for information. More and more governments around the world are defining and implementing open data strategies to enhance the three pillars of this phenomenon (transparency, participation and collaboration). Proponents of open data hope that these changes will strengthen democracy and improve the impact of government activity through increased transparency, participation and collaboration.

They believe that this will lead to greater efficiency through an information infrastructure that allows for better reuse of data. They are also motivated by the potential of open data that can be used to generate new innovations. The culture of open data is based on the availability of data in a communication-oriented way: this allows knowledge to be generated through transformation effects, where data is provided or used in applications in an innovative way.

18 Mashups: Blends.

France is one of the few countries in the world to have an organization like INSEE, which produces a lot of statistics. We therefore have a very significant potential for value-added activities based on data processing. The benefit is both private and public. It is in the interest of States to let the hundred flowers of these new activities bloom: not only do they create value directly, but they also provide services to the whole community.

The current debate around open data evokes the reflection of public entities and companies on the availability of their data, in order to generate business or spin-offs in terms of image (transparency, information, communication, etc.). In the context of open data, data can circulate freely to encourage the search for innovation or simply to increase productivity.

2.4. Developing the culture of data sharing

The increasing production of data, generated by the development of information and communication technologies (ICTs), requires greater openness and useful sharing that is becoming a powerful force for change in the economic and social world. Big Data has become an important element in the decision-making process of business leaders. They must learn to work together by innovating with their customers, taking into account their behaviors and listening to them to create the products and services of tomorrow that will enable them to generate profits to achieve their goals.

This requires us to rethink concepts of data value in a world moving towards digital interoperability, sharing and collaboration. The phenomenon of "open data" has spread around the world, thanks to its ability to generate both economic and social value. To this end, various states of the world are actively participating. The term "open data", which precisely means that all data are freely accessible, appeared at the end of the 2000s, first in the United States and then in the United Kingdom, which were the first to embark on the process.

> The President of the United States has gone very far in making the case for opening up the data. It is not simply a matter of opening the data for the sake of transparency. He said, "Entrepreneurs can take advantage of open data to create jobs by solving the problems Americans face that the government can't solve".

Box 2.13. *Open data and job creation*

It can be said that the principle of free access to data dates back to the Declaration of the Rights of Man and of the Citizens, which states that: "*society has the right to hold any public official accountable for his or her administration*". This obviously requires complete transparency of the actions of the State. This imperative, which is unfortunately always imperfect, has been given concrete expression throughout our history by the recognition of a right of access to public archives (1794), by the creation of a major public statistical information service (INSEE-1946) and by the establishment of a public service for the dissemination of law.

2.5. Storage of data in databases

[19]Computer databases are used in many companies to store, organize and analyze data. Find out everything you need to know about it: what is a database, what is it used for, how it works, what are the different categories and what are the best ones?

A database is a set of data organized for quick, manageable and up-to-date access: in a database, in rows (records), columns (fields). They correspond to a table. These data are indexed to easily find the data you are looking for using computer software. Each time new data is added, the table or database is updated.

Databases are stored as files on various digital media. Databases or tables contain multiple records (table rows) in multiple fields (or columns) that contain data.

Companies typically create multiple databases that contain multiple data to be aggregated with many independent files that include linked files or even overlapping data. The multiplication of users and bases must make it possible to work quickly and efficiently from anywhere. Thus, it is possible to carry out data analysis. The different types of databases are developed to meet these requirements: text-oriented, hierarchical, network, relational, object-oriented and so on.

19 According to the file of L. Bastien de LeBigData: https://www.lebigdata.fr/base-de-donnees.

These include:

– hierarchical database;

– network database;

– text-oriented database;

– SQL database (relational);

– distributed database;

– Cloud database;

– NoSQL database;

– object-oriented database;

– graphics-oriented database.

Box 2.14. *Different types of databases. From an article on the LeBigData platform (Bastien 2019)*

It is necessary to adapt and to try new approaches, new methods, new knowledge, and new ways of working, which leads to new properties and new issues since a logical referencing must be created and implemented. The question is no longer what data to store, but what can be done with that data?

Achieving the full potential of data depends on how it is presented, it must be used and reused in multiple ways without diminishing its value. This means making data available in the right form and at the right time to any interested party wishing to exploit and add value to it. Generating value at the different stages of the data value chain will be at the center of the intersection of openness and voluminous quantities of data. In addition, one of the main benefits of open data is that it fosters the development of a culture of information sharing and intersectoral collaboration (GTDO 2011). Increasing data volumes, processing speed and diversity pose new challenges for companies and affect their decision-making processes.

But it is imperative that these combinations be made with rigor and methodological transparency. This leads us to say that it is not a question of the size of the data but rather what can be done with it. After all, the purpose of collecting and analyzing all this data is not just to know but also to act.

Big Data processing in marketing, for example, must provide the tools to establish the mode and level of engagement necessary to attract and retain each customer at the lowest cost and to manage the ongoing relationship at an optimal level of profitability.

2.6. The appearance of buzzwords: Big, Open, Viz, etc.

In recent years, few economic articles and documents have not mentioned the terms related to the "data" family, such as "Big Data", "open data" and "data mining" when referring to the company. The result is the conviction among researchers that these concepts are emerging as new challenges that can benefit the company.

Below are two graphs on the rise of Big Data and open data.

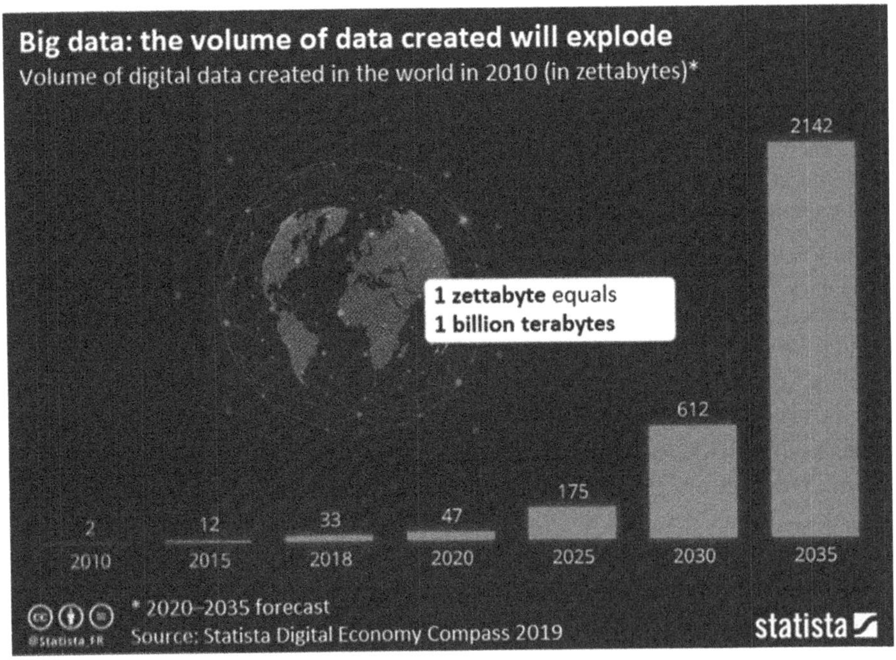

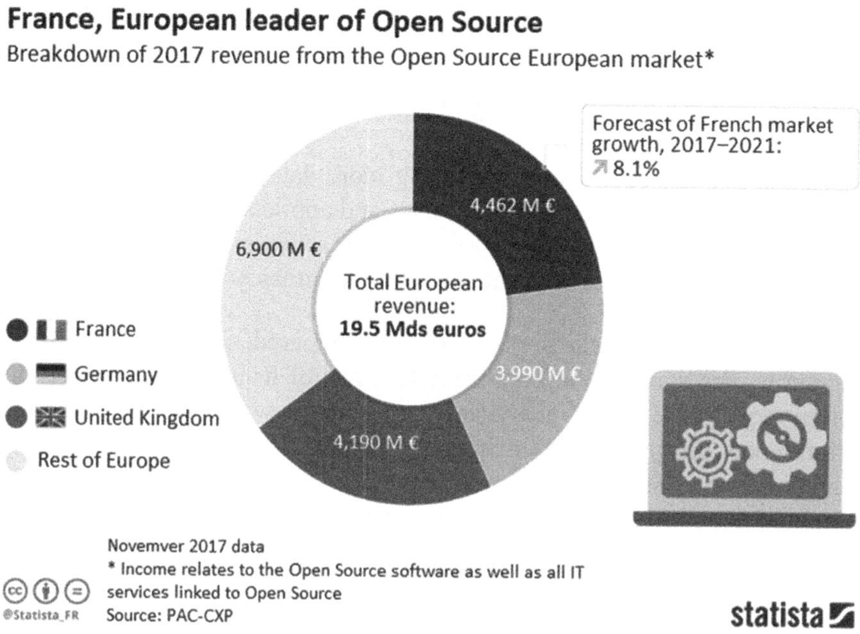

Figure 2.17. *Big Data and open data volumes (source: Statista). For a color version of this figure, see www.iste.co.uk/monino/control.zip*

2.7. Conclusion

The ultimate goal is not only to collect, combine or process all the data, but above all to increase their value and efficiency. This means moving from “Big” data quantity to “Smart” data intelligence, because the relevance of business strategies now depends on data quality.

Big Data is the essential contemporary concept that has revolutionized the way many companies work, because their horizons extend beyond borders and economic activity is part of a globalized world. Smartphones, tablets, cloud, GPS, Web, etc. are new production tools that refine the raw material: data. Data, this new strategic asset, will undoubtedly influence the overall strategy of the entity. As a result, the data that the technological base exploits have become fundamental in the redesign of decision-making processes.

It is necessary to clean, analyze, cross-reference data (structured and unstructured) produced by: companies, the State, social networks and so on to create new uses. Big Data is the general term used to describe the exponential increase in data volumes, alongside of course the growth in the ability to transfer, store and analyze this data. But storing huge amounts of data is one thing, processing it is another.

Big Data is not only about processing more data, but also about valuing data. Analytical skills to develop new uses and applications for the masses of data are therefore crucial. This is true for data scientists, who have both the technical skills to analyze data and the ability to grasp policy issues.

The quality of the data is essential: well-prepared, sorted and integrated data allows the company to take full advantage of its analysis. Without this preliminary phase, the analytical processing will not produce the performance information, which the company needs to strengthen its competitiveness. The challenges of Big Data are therefore related to the volume of data, its variety, processing speed and value.

Indeed, with the progress of the digital revolution, open data presents a common repository, which is favorable not only to the economy and society, but also to research and the propagation of knowledge. The open data movement aims to promote knowledge and innovation through information sharing and multisectoral collaboration. This movement is made possible by information and communication technologies that allow the exchange of an almost unlimited amount of data (Big Data). By opening its data to everyone, an organization allows its data to be reused for different purposes. The results of this secondary use can then be shared with the community, creating a multiplier effect.

The phenomenon of open data, which opens up digital data from administrations to the public space, developed very rapidly in the United States before spreading to England, France and many other countries. For Tim Berner Lee, the phenomenon of open data should be the future logic of Web developments. Moreover, with the development of data of all kinds, the evolution of the quantities of data should generate strong added value both in terms of use and in terms of decision-making and development aid, thanks to a better synergy of actions and actors.

Moreover, one of the main benefits of open data is that it fosters the development of a culture of information sharing and intersectoral collaboration. As a multisectoral principle, open data can generate benefits in the economic, cultural, and social spheres. The potential impact of such an opening can therefore be exponential.

> Harbulot (1992) considers that information warfare operations in the economic field can be divided into three categories:
>
> – deception, misinformation, manipulation, discredit;
>
> – counter-information, identifying the opponent's weak points, exploiting his contradictions, hitting his Achilles heels, using verifiable information;
>
> – resonance, creating networks of influence, moderating discussion forums and so on.

Box 2.15. *Economic domain in three categories*

The data collected and analyzed are transformed into information that is intended for the production of knowledge. "Data" has become a factor of competitive advantage, a real raw material for the so-called "new economy". This process was initiated in the early 1990s and marked the emergence of the modern concept of "information warfare". The informational advantage depends on the time factor. Thus, the increasing weight of open sources implies a reflection on the operational developments of economic intelligence.

3

From Data to Information: Essential Transformations

3.1. Value creation from data processing[1]

Today, while long-term economic growth depends on the accumulation of knowledge and the ability to introduce new products, processes, services, business models and organizational methods, the competitiveness of enterprises is determined by their ability to organize a beneficial environment for science, technology and innovation. In this age of "Big Data", data, which is of central importance in corporate decision-making, is a new and valuable tool to feed their strategic vision.

In an ever-changing world, data, which is increasingly voluminous, rich and varied, represents a major asset for companies, provided they know how to capture, combine and analyze it. Data can be seen as a digital asset if it has the potential to create value for companies. They can predict possible outcomes by studying structured data and an ever-expanding body of unstructured data from different sources and then sharing these results to allow for meaningful correlations.

By publishing reusable data, organizations (public/private) thus promote the development of new services, as well as the dissemination of their content. The provision of more and more "open data" is important. Open data is a vast movement for the opening up of data, initiated in the United States during "Freedom Act" and which, like Google Maps, Amazon or

1 Monino and Fechtali (2017).

eBay, has revolutionized the way we consume data. This opening will give rise to numerous applications and promote innovation with tools for developing these data.

Faced with the many challenges of opening up data, French and European public cultural institutions have initiated early publication approaches using the latest models of the "semantic web" or "data web". Data is "the new world currency", and the web is the "bureau de change". Companies have understood that good decisions require good data management. The integration of Semantic Web technologies into data management systems is one approach that brings new perspectives.

One of the challenges posed by the Internet through the web is to process and understand the increasing amounts of data and transform them into relevant information. But the design of information and its display or presentation, is a growing concern of the web professions. An intelligent Web is now the one that represents an intermediary of digital tools associated with a visualization interface. This makes it possible to make graphical representations that enable and facilitate decision-making. This is where the need for an optimal "data visualization" comes into play, adapted to a very operational exploitation of this Big Data. "Data visualization" revolutionizes the use of data and offers simple and efficient access to information from the mass of data produced.

Data accumulation is not information; it is numbers, words, events existing outside a conceptual frame of reference. In the absence of context, the individual data do not have much meaning. The data is raw information.

Openness and data sharing are the best way for both government and business to organize, communicate and unleash the universe of collective intelligence[2]. The culture of open data is based on the availability of data with a communication orientation. This allows knowledge to be generated through transformation effects where data is provided or used in applications in innovative ways.

It is also a lever for building working relationships with actors passionate about the general interest, in order to extend public action by designing new

2 Open data at its vademecum, September 2013: https://www.modernisation.gouv.fr/home/lopen-data-a-son-vade-mecum.

services that are favorable to all parties. The objective is to promote or facilitate reuse and reinterpretation in the most automated way. This represents a very important potential for value-added activities based on data processing.

In France, INSEE provides a mass of economic, demographic and sociological data (to be supplemented by those of the OECD, Eurostat, the IMF and INED). These data are not completely raw, but already processed simply. In turn, these data can be used as explanatory variables for "models" that are used by various users to make forecasts, correlations or cross-tabulations.

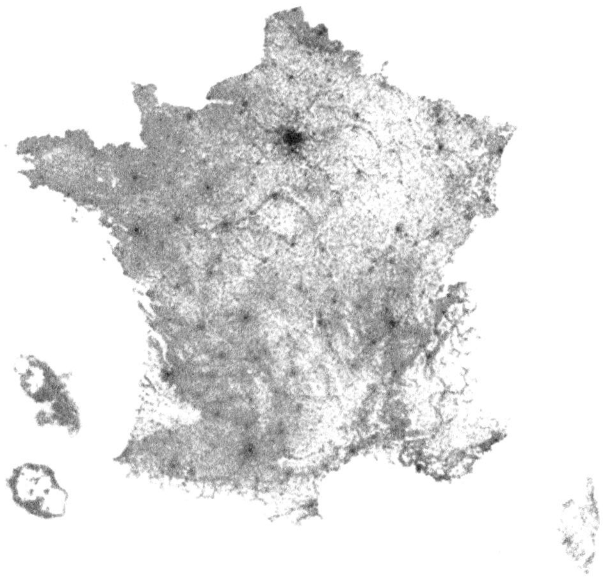

Figure 3.1. *Map of the divisions of the population. For a color version of this figure, see www.iste.co.uk/monino/control.zip*

The INSEE has put shape queues of the population: two meshes are available for downloading with a 1-kilometer step on the whole France and 200 meters on each region. This map of France provides data corresponding to population estimates made on 1 km by 1 km tiles (source: http://www.insee.fr/).

Box 3.1. *INSEE and the grid pattern*

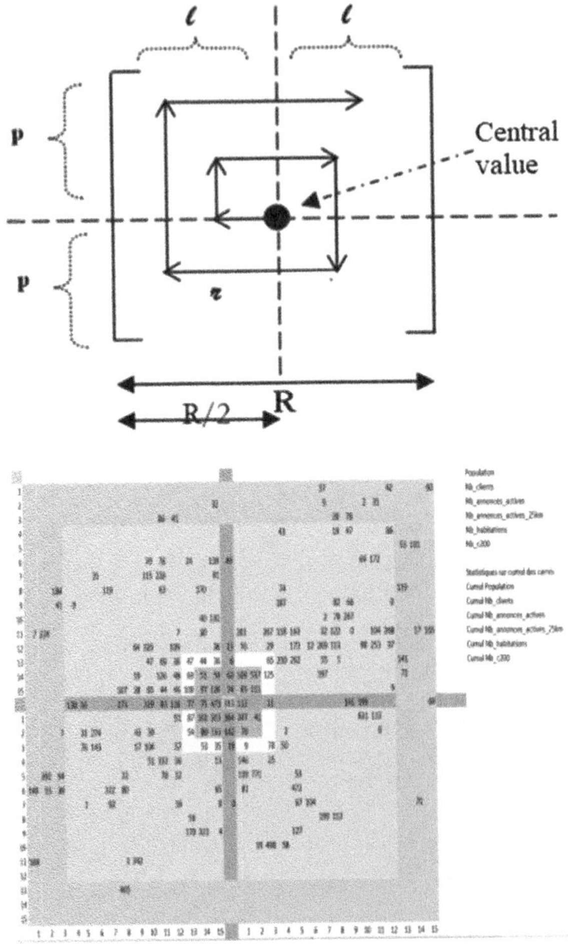

Figure 3.2. *AUTOUR.CIM. For a color version of this figure, see www.iste.co.uk/monino/control.zip*

In the same perspective, and based on INSEE data and data from Autour.com, using the Hérault department, the TRIS laboratory has developed an algorithm called: “snail algorithm”, which gives the situation or position of what we are looking for in relation to an address that we summarize to a reference point that constitutes the search point. Work done by J-L Monino for the start-up Autour.com supported by Alter-Incub in Montpellier.

Box 3.2. *A start-up population tracking application AUTOUR.CIM*

[3]The French government's strategy of opening up its data has materialized with the launch of data.gouv.fr in 2011. Similarly, on January 27, 2011, the City of Paris launched the "Paris Data" platform.

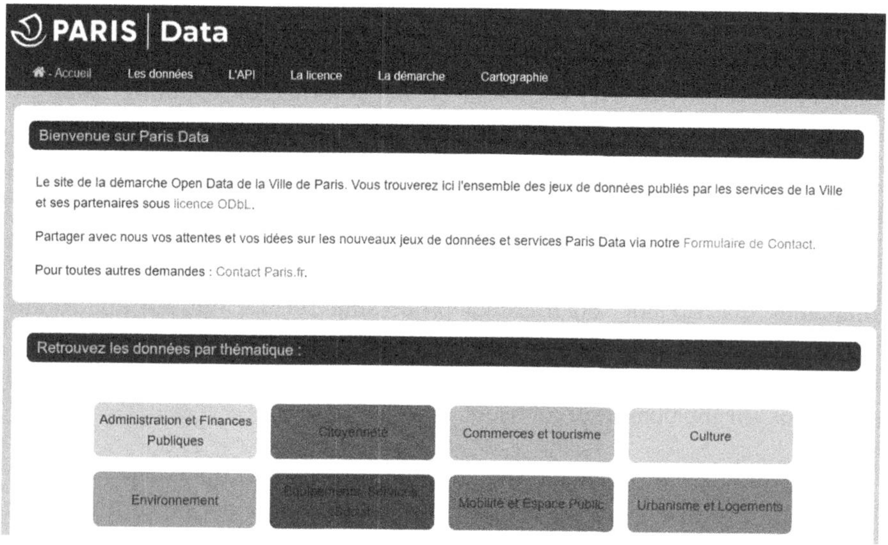

Figure 3.3. *Open data on the city of Paris. The site of the Open Data initiative of the City of Paris. You will find here all the datasets published by the City services and its partners under ODbL license. For a color version of this figure, see www.iste.co.uk/monino/control.zip*

In order to be really usable, for the opening of the data to have its full effect, the data must be processed. Not all entities to which these data may be useful have the statistical capacity to link these data to each other and draw operational conclusions. Thus, in economics, not all companies have the statistical skills to identify connections.

All econometric techniques constitute a particular body of knowledge, which may be possessed by governments and large public entities, but probably not by all enterprises. Without econometrics, there is no precise economic intelligence. This data must be processed so that it becomes "information".

---

3 The law of July 17, 1978, established the principle of a general opening of public data and granted access to public information, thus laying the foundations of the information society. The right to re-use public information was also created by the Directive.

Processing algorithms are available, which users can access and which carry out the data processing exactly as they wish. Currently, the processing chain forms a kind of "ecosystem" around the "data":

– the producer opens his or her data by making it accessible to all;

– the data user who has the processing tools and human analytical capabilities to provide and comment on the information resulting from the processing of open data;

– the end user who benefits from this information and thus develops the intelligence of his economic, political, social and even cultural "environment".

Today, the "Big Data" is a new opportunity for the creation of "online" companies that do not sell the data processing service but put algorithms online, whose use they rent to the end user, the third partner in the above ecosystem. There is enormous potential for the invention of data processing algorithms for the benefit of all users.

When an internal statistical research unit has been set up to carry out data processing, a skill is purchased for a fixed or indefinite period of time to be allocated work. If you go through a consulting firm that will collect the data, process them and draw information and knowledge from them, you are still in the purchasing process, albeit limited, but whose experts will have chosen the data to be processed and, perhaps not all the data that were available, or those that should have been chosen in order to arrive at ideally operational conclusions. With algorithm start-ups, as in all sectors, we go from buying the tool to the right to access the tool without buying.

To be more precise, in the economic field, a start-up can be created around a sales forecasting model. Let us imagine a company that sells a given product or category of products and is looking to forecast those sales. The start-up was created around the storage of a series of explanatory variables (growth rate, income growth, price index of the product category, etc.) and one or more linear regressions linking the quantity of the product sold by the company to these explanatory variables.

The end user only has to enter his own sales series and specify or let the algorithm specify the explanatory variables. The user then directly obtains the link between these explanatory variables and his or her sales, which easily leads him or her to the forecast. It did not have to go looking for external data, nor did it have to develop the "model" – already done by the start-up – and made available to all users by means of payment. There can be as many start-ups as there are models or groups of models.

We can also imagine from the coloring model (Monino and Boya 2010) that the customer is looking for the "coloring" of his communication discourse. The start-up may have a search engine for all the texts that may be of interest to the firm and may have developed a global analysis system of the coloring on all these texts, possibly based on words given by the client himself.

The coloring model provides the ability to reconcile exogenous data with changes in the price of an underlying asset and thus examine whether the coloring of the information is the same as the coloring of the stock market change. The results show that this modeling makes it possible to anticipate price changes in a statistically significant way. This model examines whether the coloring of exogenous information leads to an identical change in share price. The model for relating stock market data to informative coloring is based on Markov chains. The model has been applied in the area of financial markets; however, it is clear that it can be extended to other topics. The coloring of an exogenous event can also be of interest at the political level, in order to examine the impact of a speech on the popularity of a politician.

Box 3.3. *The coloring model proposed by Monino-Boya*

It is easy to understand the number of potential companies that can automatically process the data sources that make up the "Big Data". We are certainly facing a component of the industrial revolution that digitization has opened up.

3.2. Value creation and analysis of open databases

The OECD stresses in a 2013 report (OECD 2013) that the convergence of technological and socio-economic trends linked to the development of Internet use and the decreasing costs of managing dematerialized data leads to the generation of a large volume of data that can be exploited for the creation of new products, new wealth and new value for companies.

Sectors that create value to:
– improve research and development;
– create new products based on the data;
– optimize manufacturing processes;
– optimize targeted marketing;
– improve managerial approaches.

Box 3.4. *The OECD identifies five areas of potential value creation*

It is therefore obvious that the data must be considered as an essential intangible asset of the company. The increasing production of data, generated by the development of information and communication technologies (ICTs), requires greater openness and useful sharing, i.e. becoming a powerful force for change in the economic and social world. The birth of open data is not random, but completely linked to the Internet and its mutations. The development of digital technology and the potential uses of data call for the acceleration of this movement; more data must be shared freely, in open formats, with freedom of reuse.

Open data, or the phenomenon of "open data", has spread around the world despite its novelty, thanks to its ability to generate both economic and social values. For this reason, various states of the world are actively participating in this evolution by allowing access and reuse of their data. Open data means that the public sector is abandoning its role as a custodian of data and replacing it with a new role as a provider of data and information, leading to a realignment of power dynamics between the public and private sectors.

This phenomenon has attracted a lot of attention in recent years, and one of the central principles of the Open Data Initiative is to make public data accessible to promote innovation based on this data. A question arises: How can we take advantage of or create value by simply reusing or transforming this open data?

Data deposits are a raw material that revolves around a loop made up of the suppliers who emit the data to consumers, and who, through a transformation operation, will produce information and therefore value from the exploitation of the data. But in order to generate value, this data must be accessible and available to all.

In 2007, the Open Government Data working group defined the following eight principles of data accessibility. A data item is considered open if it is:

– complete;

– primary;

– timely;

– accessible;

– exploitable;

– non-discriminatory;

– non-owner;

– free of rights.

Box 3.5. *The eight principles of data accessibility*

In order to advance in open data, it is necessary to be able to admit this variety of opening conditions. We therefore potentially have a very large number of data concerned, very diverse in their characteristics, which can be crossed with other types of data, from social networks, for example, to enrich applications or make new ones. This is possible through analysis and data processing. Relevant data cross-referencing is the basis of all applications based on open data.

The term open data refers to public or private data that can be reused. The notion of openness of data refers to the "general public" dimension, i.e. that all persons can consult a set of data in a clear and simple manner. This is about free data that generates benefits through its reuse.

However, a theoretical framework needs to be defined that explains the impact of making data available. That's what Mr. de Vries and G. Sawyer presented for the "European Space Agency" based on[4] empirical modeling.

4 https://www.futura-sciences.com/sciences/definitions/univers-esa-2466/.

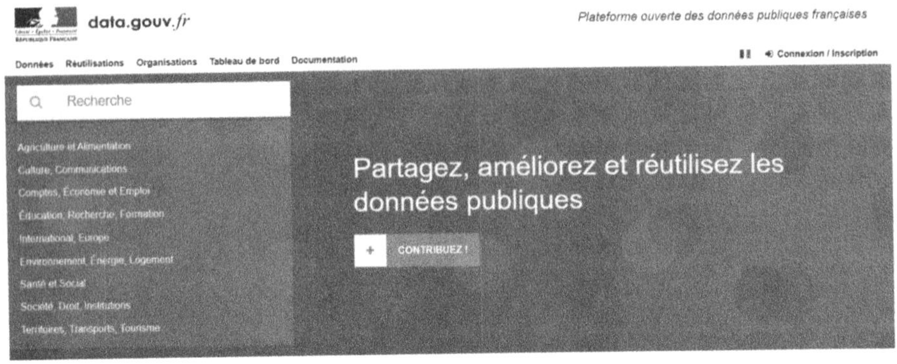

Figure 3.4. *The data.gouv.fr website. In the same perspective and in order to make the best use of open data, the Open Data portal "data.gouv.fr" in France makes available more than 350,000 datasets under the "Etalab Mission Open Licence". The data released in this way is of very diverse nature and can generate interesting applications. For a color version of this figure, see www.iste.co.uk/monino/control.zip*

This modeling identifies three phases during which the effect of data opening varies:

– a data opening phase represents a cost for the administration;

– a phase of culture where the number of re-users grows, public services become more efficient and reuse begins to generate profits for society;

– a harvest phase where the effects of openness are being felt on employment and public finance activity.

The reuse of data will allow for enriching these data, to cross them with other existing data and to offer new products or services. Open data makes it possible to create an ecosystem of partners. Opening this data is a big step towards a new stage of co-production of new activities for companies. Data becomes information, and then it becomes knowledge and constitutes knowledge, and can be a source of innovation by cross-referencing it with other data from various sources.

Faced with this reality, the challenge for the company is to manage this data value chain. In just a few years, open data has thus become a growth sector, attracting many companies and start-ups whose mission is to process, sort and analyze this data to make sense of it and make it usable. The new connected objects, watches, bracelets, the smart car, etc. will produce data that will have to be shared with those who wish to enrich or aggregate it.

Engaging in an open data approach means making a selection of raw data accessible and easily reusable by all: public, scientific, citizen and corporate data. Enriching analyses, developing or improving products and services are the main benefits sought by companies through the exploitation and reuse of open data. The decision to embark on an open data approach must be considered in light of the benefits it can bring to the company, in close connection with the objectives of its digital strategy and its innovation strategy.

Current data openness initiatives such as "data.gouv.fr", launched at the end of 2011, represent an important step in the implementation of the data web.

3.3. From data to information: the "DataViz" or data visualization

Nowadays, the various uses of the Internet, social networks, smartphones and other connected objects generate billions of data that can be stored, analyzed and exploited. The advent of high-performance processing technologies can enable systematic, real-time analysis of this immense data potential. However, the latter is increasingly complex to interpret, and it is becoming truly relevant for companies to invest in this area, which generates income and value.

First, a team of "data scientists" must identify the right datasets, access those datasets, develop a mathematical model to address them and then begin to generate ideas. Once this work is completed, the team hands the results over to top management in the form of a report or spreadsheet, which can be difficult to understand. In fact, the results are not just the numbers on the report, but they are also the decisions and assumptions that were made throughout the previous work.

These results require a debate in which, at present, many top management cannot participate because they do not have the necessary information in a comprehensible format. As a result, managers can move forward on strategic business decisions without even realizing what they have missed. Hence, Big Data must be accompanied by a powerful visualization of the processed results in order to be more successful.

In addition to the phases of collection, processing and analysis, which represent a cycle providing new information, and which must accompany the Big data to make the most of the data warehouses. It is therefore the phase that takes on greater importance. The answer to this question lies in "data visualization", which provides a more transparent, intuitive and contextual way to use data beyond the simple numbers. The need for data integration for successful Big Data analytics is therefore undoubtedly the answer.

Companies have an interest in moving towards solutions that cover analytics in its entirety, and must complete the data value chain, from data capture to presentation and dissemination. Data visualization provides a common language for executives, data scientists and top management to discuss data. These three groups generally speak "different business languages", and data visualization will replace these differences with a visual discourse that everyone understands.

William Playfair confirms this idea by pointing out that: "In matters of calculation and proportion, the surest way to strike the mind is to speak to the eyes".

Visual data analysis is based on three essential layers:

– data visualization;

– data analysis;

– data management.

Data visualization can be used both as an exploratory function to find patterns and extract knowledge from the data being exploited and as an explanatory function to clarify and illuminate the relationships between the data. Through data visualization, companies can leverage the true value of Big Data by accelerating data understanding and enable executives to take quick action and decisions on business opportunities.

As the volume and variety of data increase, data visualization becomes even more important to stimulate a collaborative dialogue. When faced with a huge amount of data, visual clustering brings together many measurement points that can help leaders better understand the relationships between the data and make more effective decisions.

3.4. From data to information: statistical processing

In the framework of regional innovation policies, data collection is carried out at the scale of the problems raised. From this point of view, local public bodies have many sources (often untapped) and constituted on the basis of project monitoring. From this perspective, it is necessary to distinguish between data, information and statistics. These three terms are often misused. In fact, these three words are very different and must be dealt with in a specific order.

3.4.1. *Phases of data processing*

This term generally covers measurements that are made on objects of individuals or any other statistical unit. They can be digital or non-digital. They may relate to a set of elements, objects or individuals, grouped in a well-defined category of great diversity. Other data formats include text, transcripts, transcripts of grades or decisions, images, computer-based information, etc.

Data collection is an important phase in any study. Once the identification of useful data has been carried out and the criteria for their quality have been analyzed, the data can be organized and transformed into information.

Data generates information to meet the following criteria:

– to organize the data;

– to define precisely, on a case-by-case basis, what needs to be retained; provide for the archiving of raw and processed data;

– to keep as much instrumental information as possible in order to keep the memory of the individual quality of the processed data (observation periods, interruptions, etc.);

– to organize *a priori*, and not *a posteriori*, the computerized constitution of the catalogue of observations and to facilitate the future use of the data by introducing additional fields to select and identify the observations;

– to establish and maintain synthetic grids to retrieve data quickly.

3.4.2. *Processing the data*

An inventory of the available data would help to clarify and identify records, texts and numbers, as well as to ensure their usefulness for the project. This inventory of data should include as much information as possible about the observation, everything necessary for processing and references to any results already obtained.

3.4.2.1. *Processing of non-digitized data*

It is a category of archived data for which processing passes through the preliminary scanning step. The need to systematically digitize these data is questionable: are they not better archived and accessible in their present form? Are they not easier to evaluate?

There are, however, cases where only digitization can ensure reliable storage and effective processing of data. It would of course be unrealistic to recommend a systematic digitization of all documents.

3.4.2.2. *Processing of digitized data*

The creation of databases, the organization of catalogues and access to raw or reduced data are inseparable from the necessary provision of the associated processing (and/or visualization, calibration, modeling) software.

3.4.2.3. *Statistical processing of data*

How can this information be presented taking into account the scientific processing aspect of the data? Comparisons, modeling of choices, simulations, etc. help to identify suitable statistical tools and to set up specific databases. More generally, the use of computers, which not so long ago was reserved for a small and well-informed public, is now accessible to a very large number of individuals. Inexpensive, easy-to-understand software allows many people to access and process large and complex digital data extremely quickly. Statistical methods, from the most summary to the most sophisticated, can therefore be used to process the quantitative and qualitative data collected. They will also allow testing models that link these data in a formal framework.

Very often, evaluation is confused or even opposed to audit, management control or simple monitoring of implementation. Evaluations refer to two main types of analysis:

– qualitative analyses;

– quantitative analyses.

The search for public policy performance requires a combination of strategic steering and permanent monitoring of its implementation. Performance is not judged solely in terms of resources allocated. It also contains the capacity of policy makers to:

– anticipate the effects of the policies for which they are responsible;

– adjust the means implemented to achieve the strategic or intermediate objectives they have set themselves;

– change the objectives without questioning the political project.

These elements enrich the evaluation approaches of decision-makers, the aim being to improve the strategic and operational management of public action. The introduction to performance asks the following questions:

– indicators;

– tools;

– methods;

– for which objectives.

In this project, we will highlight the need to consolidate and rationalize quantitative and qualitative data using grids and indicators and to develop methods and structures that will allow systematic monitoring over the long term. This information on the observation of public innovation policies should help decision-makers in their choices.

A recruitment team can identify and justify hiring goals for the sales department using a network graph from a professional networking website. This type of data visualization quickly communicates a candidate's position in the industry based on previous business relationships. With this information, the recruitment team can move with more determination and hire the best sales executives.

Box 3.6. *Data visualization*

Information presented in a visual format also tends to achieve more adherence. Reported as in the Bistro media, people retweet the infographics much more than the articles or images[5]. If a company needs to answer a question or deliver a story, well-designed data visualization can make the message more compelling and also amplify the value of the ideas to a much larger audience.

Data visualization should be seen as the result of a well thought-out process, which includes capturing quality data to enable good analytical results, and communicating these results effectively throughout the process.

A more efficient Big Data processing and analysis process must offer visualization functions at all stages of the analytical process, i.e.:

– data collection and preparation stage: combining the various data sources is a key step in the data analysis process. The more numerous and voluminous a company's data sources are, the more it needs visualization to understand how they can help make a decision;

– modeling step: allow modeling and adjust the model to the simulation according to the differences;

– deployment stage: many tools only allow visualization during the deployment phase. Visualization plays a role here, providing the necessary information.

Data visualization is a technique of data exploration and analysis. This is an excellent solution to address the increasing complexity of this available data. Data visualization is a tool that allows data to be represented visually, represented in an illustrated form that makes it more readable and understandable.

We can distinguish three main categories of data visualization[6]:

– the data visualization can be "fixed", fixed graphics, representing the targeted data;

5 Media Bistro, All Twitter, August 15, 2012: http://www.mediabistro.com/alltwitter/infographics-on-twitter_b26840.

6 http://www.jouve.com/fr/la-data-visualisation-dataviz-dans-tous-ses-etats.

– the visualization of the data can be "animated" giving the user the possibility of scriptwriting, storytelling as well as the codes of animation at the service of information;

– data visualization can be "interactive" allowing the user to "play" to help in the decision-making process.

If the objective is to make a decision, we will use an interactive "data visualization" instead, allowing a large amount of information to be displayed with a certain granularity of numbers or nuances. The explosion of computing power currently makes it possible to realize several elements. To extract important information from such volumes of data, companies need visualization tools to explore and analyze data flows interactively. Data visualization cannot be processed without addressing the interaction.

Visualization aims to produce visual representations that can be used to analyze and understand complex data. But interactivity is a key point of visualization to better understand what is being observed. This interactivity is all the more critical as the volume of data to be explored is large. However, an increase in the volume of data to be processed and displayed naturally hinders this interactivity, requiring much greater calculation effort. The volume of data that visualization faces today is a direct consequence of the exponential growth in available computing power.

Once the analysis is completed, the results must be presented in a legible manner. The aim is to enable decisions to be made by removing the complexity and being as clear as possible about the issues. The many calculated fields represent considerable volumes of data that must be able to store first and then be exploited through interactive visual analysis. Developments in computer science have given a great capacity for data collection and generation, as well as a great power in terms of visualization techniques.

Knowledge of human visual perception can help to provide practical solutions to some of the problems mentioned. Humans are endowed with a highly developed ability to visualize information, which can play a major role in their cognitive processes.

Figure 3.5. *Clouds of text or words. "The keyword cloud": it is a visual representation of the most used keywords on a website. Generally, words are displayed in larger fonts as they are used. For a color version of this figure, see www.iste.co.uk/monino/control.zip*

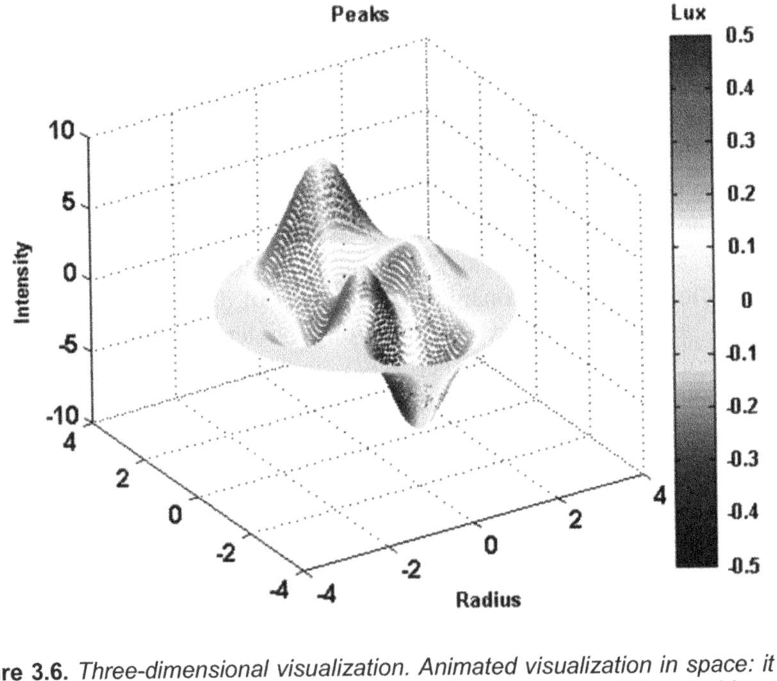

Figure 3.6. *Three-dimensional visualization. Animated visualization in space: it is an interactive image that moves so that it can be seen from different sides, while positioning the information to better read the graph. For a color version of this figure, see www.iste.co.uk/monino/control.zip*

The increase in computing power and the quality of thc graphical interfaces have made it possible to explore and develop interactive techniques that make the best use of the available data. Real-time interaction makes it easier to navigate through the data space and thus to better understand the structure and characteristics of this space; this is what characterizes the exploratory phase and improves knowledge. Subsequently, the user can formulate more efficient queries to the search system. It is at the level of interaction that the coupling between visualization and research techniques and data processing can be improved.

Interactivity is a key point of visualization. To better understand what is being observed, the company must be able to quickly change the point of view, i.e. navigate in the displayed scene, and access as quickly as possible explicit representations of the volume fields to be analyzed. Data visualization, given the amount of data to be displayed, is therefore an important analysis tool. It is becoming increasingly clear that this is an essential dimension of effective communication of results.

3.5. Turning mass data into an opportunity for innovation

With the advent of the information society from the 1980s onwards, we are witnessing the transition from a material economy to an immaterial economy. The rise of the immaterial has been directly linked to advances in scientific and technical knowledge. This development reduces the importance of physical strength and raw materials and leads to increasingly intangible products. This new "post-industrial" economy is based on intelligence, knowledge and innovation and is largely reflected in the development of services.

This gradual transformation of society organized around the production, circulation and exchange of knowledge affects both production and consumption patterns, the sources of growth and competitiveness, the organization and management of enterprises, the process of building skills and acquiring new qualifications for human capital. This confirms that knowledge is the company's most decisive resource; it is a strategic source. Looking at the knowledge pyramid, we can see that "data" is the historical pillar of knowledge.

Data is an infrastructure resource, an "input", that, in theory, can be used by an unlimited number of users to derive many applications. Within this framework, as the boundaries of economic life are changing, the most powerful companies are likely to be among those that have succeeded in exploiting this available amount of data. The ability to exploit this data can indeed make it possible to value economic activity.

The multiplication of data created by individuals, businesses and public authorities will lead to new uses and productivity gains. The magnitude of the Big Data phenomenon can be seen as an economic value in itself. Firms that are highly innovative are the most likely to rely on extensive data analysis and data mining. GAFAM or certain start-ups, revolutionaries or champions of customer experience intend to revolutionize their fields.

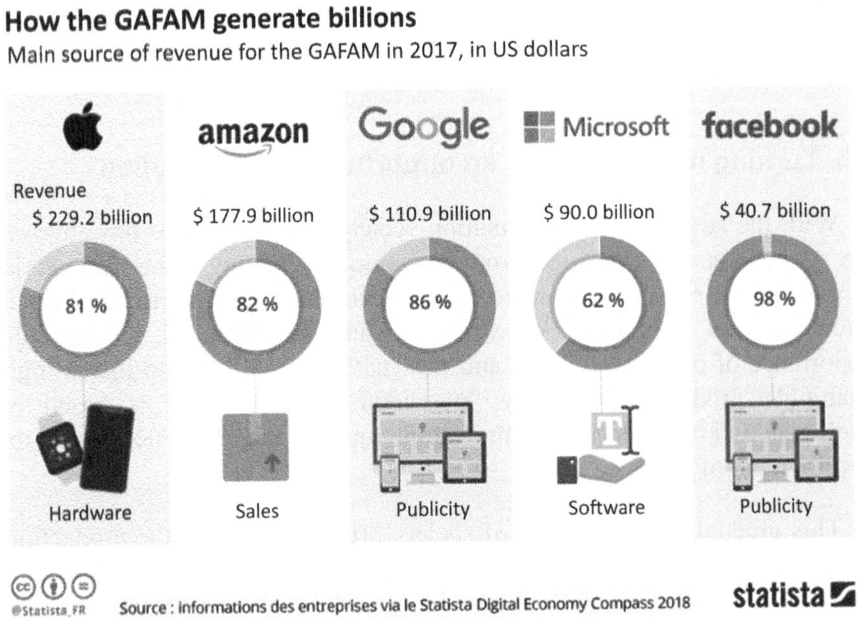

Figure 3.7. *AFSM their main sources of income. For a color version of this figure, see www.iste.co.uk/monino/control.zip*

Fast Company unveils the list of the world's most innovative companies.

Top 10 most innovative companies:

– Meituan Dianping, China's star meal delivery start-up.

– Grab, the Singaporean VTC start-up.

– The NBA, the American basketball league is being hailed for its eSport NBA 2K League.

– The Walt Disney Company, the media empire wants to launch its own streaming service to counter Netflix.

– Stitch Fix, the start-up that delivers custom-made clothing to its customers.

– Sweetgreen, the American chain of healthy restaurants.

– Apeel Sciences, a solution to keep fruits and vegetables fresh for longer.

– Square, the American Fintech.

– Oatly, the Swedish FoodTech that offers alternatives to animal milk.

– Twitch, the platform that makes TV look uncool.

Box 3.7. *The 10 most innovative firms*

The success of large companies such as Amazon, Google, Facebook, etc. provides evidence of the emergence of a fourth factor of production in today's hyper-connected world. In addition to raw materials, labor and capital, data has undoubtedly become an essential element of competitive differentiation. But their mass production of data on the Internet and in several digital forms heralds a new era of Big Data, as do the new technologies that enable its analysis.

For network companies in particular, the use of data enables unprecedented business optimization (possibility of preventing network failures, service interruptions, subscription cancelations, etc.), the development of "intelligent services" and the use of data by third parties. On this last point, by way of example, the three largest telecom operators in Great Britain have set up a joint venture, "Weve", to sell their customers' anonymized data (purchase data, geolocation data, Internet data, etc.) to advertisers.

The Boston Consulting Group also found that companies that are leaders in leveraging large volumes of data generate 12% more revenue than other companies that are not experimenting or trying to gain value from Big Data. Firms that analyze and process data in bulk (internal/external, structured/unstructured) are much more likely to be innovative.

In order for companies to take advantage of Big Data and extract value, it is necessary to:

– study the data. The company has a considerable amount of data from several sources and must understand the relationships between them;

– process the data. Thus, the company must analyze its data using a system with processing power to extract value from it;

– transform the data. In this stage, the company must use tools capable of transforming them into useful and readable information.

Big Data allows the collection of previously unknown but useful data to provide new ideas that can lead to new discoveries and thus feed the cycle of research and innovation. Big Data has many applications and is a factor in strengthening the capacity for innovation within companies by playing on the two dynamics of exploration and exploitation. The innovation will certainly come from cross-breeding and treatments that were not originally thought of.

Mining large amounts of data makes it possible to launch new products and services, new processes and even new business models. By making the data talk and giving it prominence, each company will have access to a better understanding of the context and its environment.

The collection and use of data on connected products or consumer behavior enables the company to improve its operational functioning and to forecast market trends. French software start-ups that develop algorithms help companies to optimize their stock levels on a daily basis.

Figure 3.8. *Ranking of the most attractive metropolises in the world. For a color version of this figure, see www.iste.co.uk/monino/control.zip*

In terms of "urban data", there are as many cities as there are models. Many large-scale urban innovation projects focus on making infrastructure "smart" by integrating sensors and increasing network capacity.

> As proof, the most advanced cities in the world today in this field: New York, Paris, Tokyo, Singapore, Shanghai and London have all been able to implement an original strategy, taking into account their historical, demographic and urban planning specificities.

Box 3.8. *Ranking of the world's most attractive cities*

Big Data technologies can help cities meet increasingly pressing challenges. Below are the most attractive cities in the world.

Figure 3.9. *The Intelligent City – Montpellier. Montpellier Méditerranée Métropole, true to its spirit of innovation, is experimenting with the "Smart City" concept in partnership with IBM, the University of Montpellier, IDATE and several local companies. For a color version of this figure, see www.iste.co.uk/monino/control.zip*

The "smart city" is also spreading thanks to the opening up of public data and the implementation of an ecosystem of developers, creators of mobile applications.

Open data makes it possible to promote innovation approaches, open, on the one hand, between companies or public entities holding exploitable data, and, on the other hand, start-ups or SME suppliers of innovative technologies. This represents an opportunity for radically new solutions to emerge for all players.

Big Data is therefore part of an "open innovation" approach that encourages organizations to collaborate with other players outside the firm's traditional boundaries, on strategic or unexpected "open innovation outside-in" subjects, and to think about how to make the most of projects and the data produced, which the company does not plan to capitalize on directly in its core "open innovation inside-out" business.

Beyond a gigantic volume, it is the diversity of data sources that gives Big Data its full scope. However, the development of new products and services and their adaptation to uses are facilitated by the cross-referencing of a large set of data. Big Data marks a major turning point in the exploitation of data and represents a formidable lever for growth and profitability. A global understanding of a company's data and its potential, as well as the control of operating resources can become a new performance vector.

Every company in the world is swamped with data but does not always know how to use it in a meaningful way. For data to be used for innovative opportunities, large computing and storage capacities are required. Storing data online makes it accessible without any time or place constraints and is the first layer of technology that is indispensable for Big Data processing. Another indispensable factor in this new situation, and one which makes it possible to make the data speak for itself, is the power of formulas and calculation. Data flow processing and analysis tools are becoming more powerful every day (e.g. Cloud).

The Big Data revolution, which is extended by the potential of Machine Learning, is an opportunity to change things. The Big Data thus makes it possible to apprehend the heterogeneities of the sets and to adjust the analysis according to the data in real-time, to allow a more appropriate response. The disruptive nature of the innovation introduced by the spread of Big Data in an ever-widening range of industrial fields finally opens up significant opportunities for value creation and differentiation. Sectors such as insurance, retail, automotive and energy are experiencing major changes in their value chain as a result of Big Data. This value is multiple (monetary value, use value, reuse value, etc.) and depends on who is reusing the data.

3.6. Development of company assets in the web of data

Data, both structured and unstructured, which has grown exponentially in volume, nature and variety within companies, is all the more valuable because it interacts with the huge mass of data, most of which comes from the Web. And if we add to the "Big Data" phenomenon we have already discussed, the "open data" phenomenon, which consists of making data accessible, or the "linked data" phenomenon, which makes it possible to interconnect data from different sources, then the possibilities are even greater. The Web promises to be a "gold mine" for businesses.

The Web has become a privileged place to search for data; it has favored the online availability of many types of data, potentially useful for the company's monitoring system. The Web has revolutionized the way data is captured, accessed, integrated and used. Indeed, a large amount of data available on the Internet potentially contains information useful for decision-making. The growing volume and heterogeneity of information poses real obstacles that must be overcome in order to effectively exploit this gold mine.

When we use a search engine, it does not understand our query. It finds the pages where the keywords mentioned in our request appear but does not guarantee that they really answer them. Various web-based search and data collection techniques were proposed to build tools to refine the search for relevant results. The aim is to build a space for data exchange between machines, allowing access to a very large volume of data, and providing the means to manage it.

In order to understand the concept and value of data, it is important to consider current mechanisms for exchanging and reusing data on the Web. It is therefore a question of achieving an "intelligent" web, where data would no longer be stored but "understood" by computers in order to provide the user with what he or she is really looking for. The result is a global data space called the "Data Web" or "Semantic Web", which Tim Berners-Lee began talking about between 1991 and 2001. The concept was taken up and developed by the World Wide Web Consortium (W3C).

The Semantic Web offers a new platform for more intelligent content management through its ability to manipulate resources based on their semantics. A machine can handle the volume of data available on the Web and thus provide more help to people, provided it is given a certain "intelligence". In fact, the integration of semantics into the Web is not a new idea but, in fact, was born with the Web (Falquet and Mottaz 2001). The Semantic Web is the starting point for the development of intelligent Web services. Starting from the observation that these two architectures are dedicated solely to human–human or human–machine interactions, the initiators of the Semantic

Web defend the idea that the next generation of the Web must promote access to resources so that they can be processed automatically by software agents, i.e. machines[7]. To provide automatic processing capability, the Semantic Web adds a layer of metadata to the data and makes it usable by computers. This metadata will provide unambiguous semantics to automate processing. The architecture of the semantic web is based on a pyramid of languages proposed by Tim Berners-Lee, to represent knowledge on the web while satisfying the criteria of standardization, interoperability and flexibility.

In the world of the semantic web, it is not enough to just put data on the web, it is above all necessary to create links between them, so that this web of data can be explored. Linked data is a sub-project of the semantic web, through which we can find, from a piece of data, those that are related. For Tim Berners-Lee, linked data is a very important change, and for this reason, he sets out four basic principles[8]:

– use Uniform Resource Identifiers (URIs) instead of URLs to identify what you want to make available on the web as data resources;

– use HTTP addresses to enable location and access to their content;

– provide useful information about the resource (URI) when it is consulted;

– include links to other URIs related to the resource being consulted in order to enhance the discovery of useful information.

The data link allows navigation and sharing of information between different sites. Connecting a database to the Web is like setting up a gateway between a web server and a database management system. This gateway leads to deal with techniques associated with the world of computer science related to "client/server" architecture, involving web-oriented programming languages, communication protocols and data manipulation languages. Large quantities of structured data have been published, particularly in the framework of the "Linking Open Data"[9] project.

The interest of linked data lies in their ability to identify and type relationships between entities, using Web standards that are

7 http://www.cs.man.ac.uk/~horrocks/Slides/ed02.pdf.

8 http://pro.01net.com/editorial/506931/le-web-intelligent-prend-forme/.

9 http://esw.w3.org/topic/SweoIG/TaskForces/CommunityProjects/LinkingOpenData.

machine-readable and can reveal interconnections that would be impossible to produce by human intervention alone. Linked data refers to a set of best practices for publishing and linking structured data on the Web. Linked data is a set of design principles for sharing machine-readable data on the Web for use by public administrations, businesses and citizens[10].

The notion of linked data corresponds to a more technological vision of data in relation to the Semantic Web. It is a collection of datasets published using the languages of the Semantic Web and the links between datasets, allowing several datasets to be queried in the same query. Many tools are beginning to appear to store data, visualize them in various ways, offer visualization interfaces with maps, etc.

The development of semantic technologies and the evolution of data to the Web have led to changes in the representation of terminologies and metadata formats and to new possibilities for enabling machines to reason on the basis of terminologies. These metadata bases are becoming more numerous every day and can be used remotely, directly via web services. Web of Data technologies present the immense prospect of sharing, pooling and leveraging metadata on the Web.

In companies today, applications derived from the semantic web are essentially represented through software tools for data search and structuring. We can notably mention "Exalead" and its "Cloud View" platform, which enables the collection, analysis and structuring of large volumes of data (emails, social networks, blogs, etc.) from fixed or mobile systems and terminals, as well as from the web. Its objective is to enable companies to tailor their information needs to their specific issues and sectors of activity.

The term Semantic Web or Data Web refers to the vision of the Web of the future as a vast resource exchange space supporting collaboration between humans and machines in order to more efficiently exploit the large volumes of data and various services available on the Web. This new generation of web represents a double dream of its creator "Tim Berners-Lee". On the one hand, the web is becoming a very powerful means

10 EC ISA Case Study: How Linked Data is transforming eGovernment.

of cooperation between human beings, and on the other hand, it is also a means of cooperation of "machine" computers[11].

3.7. Conclusion

Currently, in a context of globalization, actors in their territories are confronted with new threats and opportunities whose control depends on the production of knowledge and know-how. In order to remain competitive, they must organize and structure themselves according to the new information resources, which are considered decisive in knowledge societies. Innovation is not a concept that necessarily implies intrinsic novelty, and what is new in innovation is not the object in question or its content, but its introduction into a given environment.

Innovation represents the source of value creation in Schumpeter's economic theory, involving new combinations of resources, new production methods and new products and services, which in turn lead to the transformation of markets and industries, thereby increasing value. Innovation is not a concept that necessarily implies intrinsic novelty, and what is new in innovation is not the object in question or its content, but its introduction into a given environment. Although innovations are often developed to solve technical problems or to address challenges, other innovations relate to the development of service processes or changes in the way knowledge is produced and communicated.

On the other hand, we are witnessing the birth of practices that propose to seek the right information, to check the right sources and to disseminate it in order to create added value. It is about strategic intelligence and/or business intelligence. The stakes of economic intelligence are imposed on all players who want to improve their level of performance on the global market. Its practice has become an obligation not to suffer, guaranteeing its economic wealth, and to learn to anticipate with a view to creating wealth. To do this, you need to know the available secure solutions that allow us: to store, exchange, share, back up and manage this information while not neglecting the legal and organizational aspects.

---

11 Tim Berners Lee, in an interview with the UNESCO Courier in 2000.

4

Information: Contextualized and Materialized Data

"Big Data", "open data", "data mining", "data management" and all other such terms have in common "the data" or earlier terms related to the "data" family. They are everywhere, and we produce them every day in our daily lives, and those with the increasing use of social networks, the Internet. These data are carried by smartphones, tablets, etc., and are accessible in a more or less simple way by companies or individuals. These terms refer to the way in which companies are led to manage the data they have or can dispose of on their customers, prospects, subscribers, suppliers, etc. In short, on all the parties with which they interact.

According to Henri Maler and Mathias Reymond[1], "The right to information rests on a single foundation: information is a public good that must be accessible to all and must not exclude any area of economic, social and political life. The exercise of this right does not only consist in the right to be informed, but also in the right to inform that journalists demand"[2].

The media and journalists are no longer the only ones with effective ways of relaying messages – just think of all those permanently posted on social networks such as Facebook, Twitter and Snapchat. However,

1 http://www.toupie.org/Dictionnaire/Information.htm.

2 Médias et mobilisations sociales, 2007.

this multiplication of statements of all kinds, coming from a variety of sources, raises a simple question whose answer is less obvious than it seems: what is information? The word in itself, in its media sense, refers to facts that have come to the public's attention. But to be considered as such, an "info" must meet at least three criteria.

4.1. What is information?

Information is data that has been recorded, classified, organized, linked and interpreted within the context of a specific study. The exploitation of the collected data requires sorting, verification, processing, analysis and synthesis. It is through this process that the raw data collected during the research will be transformed into information. This data processing process provides accessible information for decision-making.

4.1.1. *How can we define information?*[3]

The concept of information for Anne Mayère (1990) is based on its characteristics, "it is that economic analyses on information seemed to us to stop along the way, and not to allow us to fully take the measure of the phenomena at work and the related issues".

Mayère continues "Most analyses consider information as 'data' in every sense of the word, in the most mathematical sense of a processing object and in the sense of immediate and certain accessibility".

As Mayère signals, if you push this term, this definition of information data excludes information from the field of the goods. Indeed, it is not possible to control its distribution insofar as there is no constraint thwarting its access, and therefore, it cannot be appropriated and exchanged according to the principles of the goods.

3 https://www.cnrtl.fr/definition/information.

Figure 4.1. *Advice, information, rumor. According to Agathe Dahyot/Le Monde, our translation of original figure (Les Décodeurs 2017). For a color version of this figure, see www.iste.co.uk/monino/control.zip*

The multiplication of statements of all kinds, coming from various sources, raises a simple question whose answer is less obvious than it seems.

> Meaning 1: Information is the action of informing, being informed, giving “knowledge” of a fact or seeking it.
>
> Meaning 2: Information is news, details, documentation about something or someone that is brought to someone’s attention.
>
> Meaning 3: Information is current affairs, news disseminated by the media: press, radio, television, the Internet. Familiar: the news.
>
> Meaning 4: In French law, a judicial inquiry is a judicial investigation triggered by the public prosecutor in complex cases and entrusted to an investigating judge, prior to judgment. Synonym: instruction.
>
> Meaning 5: In computing and telecommunications, information is an element of knowledge (voice, data, image) that can be stored, processed or transmitted using a medium and a standardized coding method.
>
> Meaning 6: In mathematics, information theory is the discipline that studies the different ways of processing, transmitting and receiving information that are involved in any message (written, oral, computer, etc.).

Box 4.1. *Different meanings of information*

Information is a set of data contextualized and materialized by a support in order to formalize exchanges between individuals or between machines. These supports can be paper or digital documents, images, videos, soundtracks, microfilms, informal discussions from person to person and so on.

The information flow is a conceptual representation of all the information movements existing for an organization, regardless of their formats, nature, value, issues and origins (internal or external).

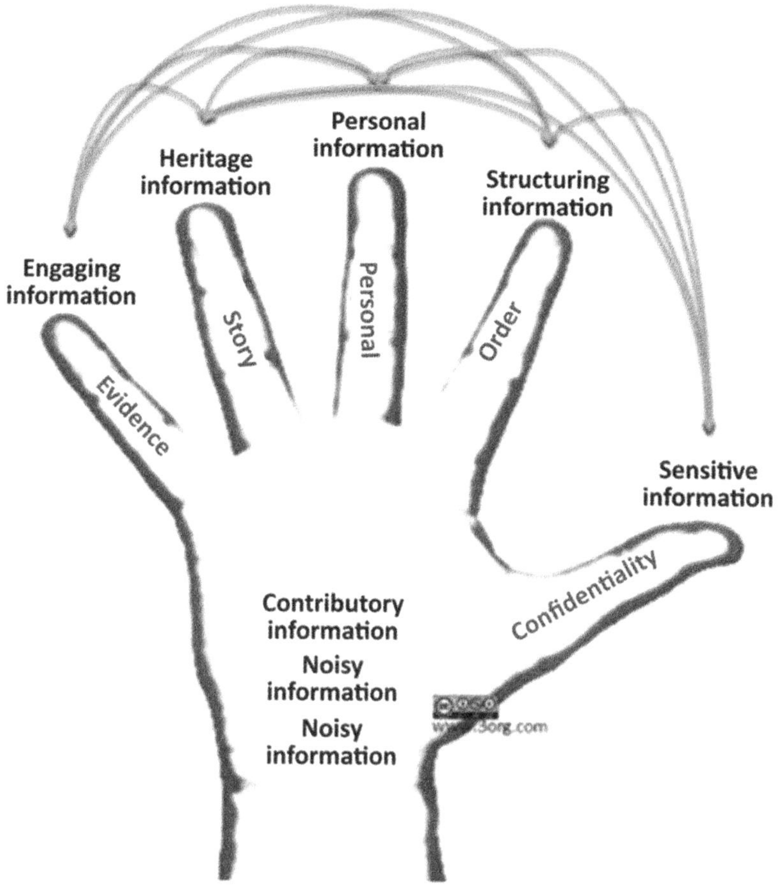

Figure 4.2. *According to Jean-Pascal Perrein, the information flow is represented by seven families of information[4]: engaging, heritage, structuring, sensitive, personal, contributory and noisyFor a color version of this figure, see www.iste.co.uk/monino/control.zip*

Companies need to move towards information governance, but information governance is often focused on risk and regulatory management.

The fundamentals are documents or sets of information structuring the uses and behaviors of uses. They contain rules, guides, indicators and all structural elements that help improve the control of the flow of information.

4 https://www.3org.com/news/gouvernance_de_linformation/les-7-familles-dinformations/.

Information governance is the set of processes, fundamentals and structures that enable an organization to better control and enhance its information flow.

Specific information governance tools are technical systems that promote the creation and maintenance of the fundamentals and enable the organization to master all or part of the information families. The tools that disseminate this governance are those that process information; they are inseparable as part of an overall approach to information governance.

The global approach to information governance is therefore faced with the management of their:

– data processing;

– storage;

– backups;

– security;

– protection.

Box 4.2. *Overall approach to information governance*

According to Taylor (1980), the value of information begins with a piece of data, which acquires value as it evolves to achieve the objective of expressing an action to make a decision. It is considered as an interpreted data flow, whereas knowledge is created and organized from this information flow. This exploitation gives even more value to the information, which must be protected. Within this framework, we can propose a knowledge approach as an understood information modeling.

A data is the representation of a symbol or symbols of what has happened or is happening at the moment. It is not significant in itself; it only becomes significant when someone interprets it and gives it meaning. It becomes information for the interpreter. But the question is how to make the information useful (Monino 2016)? The answer to this question is based on a distinction between three types of information:

– white information: easily and legally accessible information (media, trade events);

– gray information: information that is legally accessible, but characterized by difficulties in knowing of its existence or in accessing it (database, expert interviews, extranational markets, etc.);

– black information: restricted information whose access or use is explicitly protected; it actually requires authorization (partnerships, syndications) to be legally obtained.

Information is a message with a higher level of meaning, which requires a medium, a sender and a receiver. It is also an object susceptible to a wide range of treatments and manipulations through multiple supports. It is therefore a raw data that the subject then transforms into knowledge through a cognitive or intellectual operation. Whereas knowledge is the extension of information to which one adds value because it is based on intention. As such, it is a specifically human phenomenon that concerns thinking in its active and contextual relational frameworks. Each piece of knowledge is part of a process in which other knowledge is already present. It originates in the brains of individuals and is built from information that is transformed and enriched by personal experience.

Information is data that has been recorded, classified, organized, linked and interpreted within the context of a specific study. This data processing provides the information available for decision-making. In general, there are several kinds of information that can be summarized in Figure 4.3.

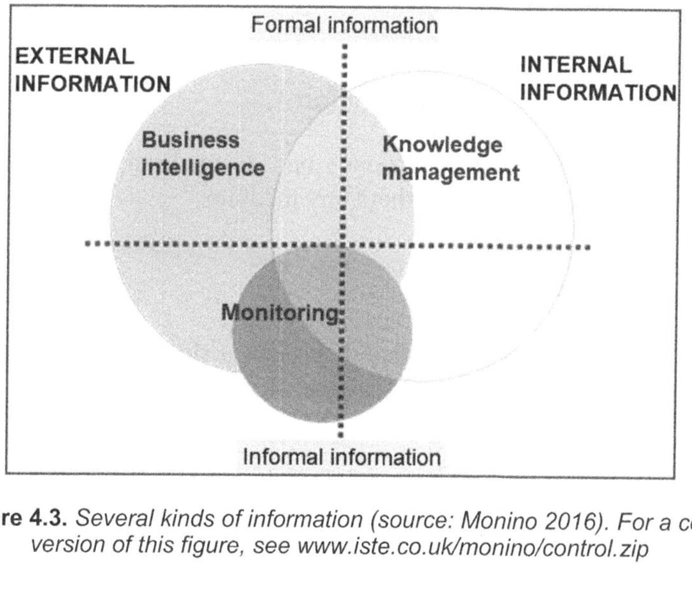

Figure 4.3. *Several kinds of information (source: Monino 2016). For a color version of this figure, see www.iste.co.uk/monino/control.zip*

4.2. Internal and external information

Information refers to data that has been recorded, classified, organized, linked and interpreted within the context of a specific study. These data processes provide information for decision-making. In general, there are several kinds of information.

4.2.1. *Internal information*

It provides information on the actual state of the organization (a company, for example), based on tangible knowledge such as:

– the capacity of the machine park, the financial situation and cash flow, the organization, the order book, etc.;

– know-how, human resource situation, customer relations, etc.

4.2.2. External information

It brings knowledge from a thematic watch such as:

– standards, patents, regulations, products and processes, customers, competitors, mergers and acquisitions;

– vision of the future, trends, market forecasts, foresight, political and social developments, etc.

4.3. Formal and informal information

Formal communication is done on a medium, while informal communication is done orally without any medium.

4.3.1. *Formal information*

Formal information is data that has been written or distributed on a medium (paper, multimedia, sound, image, etc.).

4.3.2. *Informal information*

Informal information is data collected from an oral source or that has not been explicitly formatted for publication.

4.4. *Importance of information*

Accumulation of information is not knowledge, but it is the set of data processed and transformed that becomes information, and validated and confronted that begins to make sense.

4.4.1. *White information*

White information is easily and legally accessible (media, commercial events, etc.)

4.4.2. *Gray information*

Gray information is legally accessible, but characterized by difficulties in knowing its existence or in accessing it (database, expert interviews, etc.).

4.4.3. *Black information*

Black information is information that is restricted and whose access or use is explicitly protected and requires authorization to be legally obtained.

These three concepts should now be of interest to operational decision-makers (marketing managers, financial managers, etc.) when it comes to analyzing, in real-time, the immense potential of data held by companies.

To meet the challenge of "Big Data", measures should be taken that include all the tools for processing the data, more restrictively all the actors who analyze these data. This will only be possible through an awareness of the gains offered by "data enhancement". The data in the databases organized, reorganized and processed by statistical methods, or econometric modeling, become knowledge.

Information brings something that must concern the public for whom it is intended.

One of the principles of the use of information is not to give information without first verifying it with, as the case may be, the persons directly concerned, several witnesses or experts, etc.

"I want the right information at the right time!" For that you need information, not advice! It must be:

– defining objectives;

– research and data collection;

– factual;

– verified;

– sorting and storing data;

– having relevant information.

4.5. Décodex[5] set up by Le Monde

Décodex, launched in early 2017 by Le Monde, is a tool that aims to fight against the dissemination of false information. It helps Internet users find their way through the crowd of sites that produce or relay false information.

The Décodex allows us to share information; it allows the provision of different tools to assess the reliability of the site on which it is hosted, we have two extensions:

– a Chrome extension;

– a Firefox extension.

5 Against the dissemination of false information, Le Monde launched Décodex (Les Décodeurs 2017).

Décodex tells you, in real-time, during your navigation, if the site is rather reliable or if it regularly broadcasts false information. The particularities:

– a search engine on the site of Le Monde;

– a verification of the reliability of a site;

– pedagogical articles, especially for teachers;

– tips on how to tell the difference between information and an information source;

– tips for verifying information, images, videos on the Internet, etc.

Box 4.3. *The DÉCODEX set up by Le Monde*

4.6. Conclusion

How can this information be presented taking into account the scientific processing aspect of the data? Comparisons, modeling of choices and simulations, among others, contribute to the identification of suitable statistical tools and, more generally, to the constitution of specific databases. The use of computers, which not so long ago was reserved for a small and informed public, is today accessible to a very large number of individuals. Inexpensive, easy-to-understand software allows many people to access and process large and complex digital data extremely quickly. Statistical methods, from the most summary to the most sophisticated, can therefore be used to process the quantitative and qualitative data collected. They will also allow testing models that link these data in a formal framework. Quite often, evaluation is confused or even opposed to audit, management control or implementation monitoring. The evaluations actually refer to two main types of analysis, qualitative and quantitative, which they cross-reference with a certain amount of subjectivity, for example, the search for performance induced in public policy engineering and the articulation of strategic management with the permanent monitoring of its execution. Performance is not only judged in terms of resources allocated; it also includes the ability of policy makers to:

– anticipate the effects of the policies for which they are responsible;

– adjust the means implemented to achieve the strategic or intermediate objectives they have set themselves;

– change the objectives without questioning the political project.

"We have to change the culture of the company!" The decision-making system is not a fixed system. It has to adapt and evolve over time, and to do this, it must:

– share information;

– evaluate the quality and relevance of decisions;

– question itself to reformulate, change and improve.

5

From Information to Knowledge: Valuing and Innovating

Companies are very aware of the importance of knowledge and even more so of how to "manage", enrich and benefit from it. Beyond all the factors (financial, technical and other), the knowledge that the company possesses also represents an important survival factor, whether it is market knowledge, legal, technological, normative or other information.

Knowledge is the extension of information, to which value is added because it is based on intention. It is a specifically human phenomenon that concerns thinking in its active and contextual relational frameworks. It represents an acquisition of information that is translated by the human factor and requires analysis and synthesis to assimilate, integrate, criticize and admit new knowledge.

From information to knowledge. "How can I make the information useful?" Once the overall objectives in terms of information have been set, and the research, collection, sorting and storage missions have been validated, it is necessary to:

– analyze the information;

– use the results to highlight aspects that assist in decision-making.

5.1. Innovation as a driving force of growth

Innovation has received increasing attention from theorists who consider it as a determining factor in the success of a company in its market.

Innovation is a relatively old concept first formalized by Schumpeter. His work is mainly known for the development theory he designed. It is within this framework that the author has presented a true theory of innovation. Schumpeter considers that innovations are designed during periods of crisis, before they themselves fuel growth. It defines innovation as: "the successful introduction of a new product, a new manufacturing process or a new organizational form of the company to the market".

5.1.1. *Innovation and the intangible economy*

This view implicitly refers to the destructive/creative consequences of new forms of production or organization, at a microeconomic level, whose effects may be visible at the macroeconomic level (growth). Indeed, for him, innovation becomes the result of routine R&D work carried out in large industrial research laboratories. He considers that it is teams with very specialized skills and carrying out a very routine action that are at the origin of innovation. Definitions of innovation vary depending on the context in which it is used. Thus, Bayerre (1980) proposes to consider three contexts in which the term innovation is used:

- the overall process of creation;
- the adoption of a novelty by a company;
- the novelty itself.

Each of these approaches has its own definition of innovation since they are on different levels. The first refers to the creation of innovation, the second to its use and the third to innovation as the result of maturation without concern for its implementation. Innovation, in this sense, is no longer the exception, it is becoming the rule and has become a permanent feature of business life (Rothwell and Dodgson 1994; Cohen 1995), and it is an interactive process involving many actors in companies and their environment.

According to the OECD, "the minimum criterion for a change in a firm's products or functions to be considered an innovation is that it must be 'new to the firm' (or result in a significant improvement)". This includes products, processes and methods that firms are the first to develop and those that they have imported from other firms or organizations. With the advent of the information society, from the 1980s onwards, there was a shift from a material to an immaterial economy. The rise of the immaterial has been directly linked to advances in scientific and technical knowledge. This development reduces the importance of physical strength and raw materials and leads to increasingly intangible products. This new economy, described as “post-industrial”, is based on intelligence, knowledge and innovation and is largely reflected in the development of services.

The emergence of the immaterial has led to the development of:

– research activities through the practice of innovation;

– faster and more reliable information circuits;

– staff skills (good training and motivation of its staff to better serve the client).

This gradual transformation of society organized around the production, circulation and exchange of knowledge affects at the same time the modes of production and consumption, the sources of growth and competitiveness, the modes of organization and management of enterprises, the process of building skills and acquiring new qualifications for human capital. Since then, Drucker has identified knowledge as the new basis for competitiveness in post-capitalist society. This confirms that knowledge is the company’s most decisive resource and a strategic source. It presents the firm as a dynamic, evolutionary and almost autonomous system of knowledge production and application.

In 2006, Maurice Lévy and Jean-Pierre Jouyet drew up a report on the situation in France with regard to the opportunities offered by the immaterial economy. According to the authors, immateriality is today the key factor for growth. The text published by the Commission, in 160 very dense pages and in a pedagogical manner, sweeps over a vast field, from the analysis of macroeconomic developments to the formulation of 68 proposals, which are both concrete and disparate. Much emphasis is placed on the shortcomings or failures of state action.

The report is divided into three parts: the first is based on knowledge to provide France with a better framework for the emergence of talent in schools, universities, research centers, the cultural sector and businesses; the second denounces regulation as a means of putting in place rules for the operation of markets and the third contains 68 proposals advocating a threefold strategy for change: reflexes, scale, with emphasis on the European dimension, and a model for adapting the French economy to the new model of the intangible. Infrastructures are essential to enable France to acquire new networks (digital, very high speed, information and knowledge).

5.2. Knowledge: the key to innovation

Knowledge occupies an interesting place in the innovation process, the relationship between knowledge and innovation has long been recognized, and when we look at these relationships, we are surprised to find that the two notions are linked. Innovation is the creation and application of new knowledge to make it productive (Drucker 1993). Consequently, the current trend, in line with evolutionary work, is to view innovation as a phenomenon of knowledge and skill creation and thus learning. This knowledge-based approach requires companies to develop their skills, to learn more easily and quickly in order to be able to respond to the rapid and complex changes in their environment. This learning can be fostered by the development of new information and communication technologies (ICTs), which are tools that enable companies to adapt more quickly, while at the same time they represent constraints that force these same companies to evolve even faster.

Moreover, the contemporary period is characterized both by the greater strategic importance of knowledge as a source of competitiveness for companies and by greater opportunities to combine knowledge to produce new knowledge. They have an increased need to have access to new knowledge and to be able to use it for their own purposes.

Combining knowledge to produce new knowledge requires investments which themselves come from knowledge for a double purpose:

– the direct effect is to broaden their own knowledge base to increase their competitiveness;

– the indirect effect of increasing their ability to seek potentially useful knowledge from knowledge produced outside of them.

Box 5.1. *Knowledge and new knowledge*

In this context, the company must call upon external knowledge, which it does not directly control, and articulate it with its own knowledge base.

The work of Abramovitz and David (1996) provides a good illustration of increased investment in knowledge. Drucker (1968), who was one of the first to detect the advent of the information society, traced this craze for knowledge, with the institute now as the main factor of production. Foray (2000) considers that knowledge-based economies are based on a twofold phenomenon:

– a long trend of increasing resources devoted to the production and transmission of knowledge (education, training, R&D, economic coordination);

– a technological event that would be the advent of new information and communication technologies.

It is the encounter between these two phenomena that plays a primordial role in changing the movement of knowledge (production, transmission), which will be directly reflected in the economy. Through ICTs, it is not only the quantity of knowledge and its speed of circulation that evolves and intensifies but also the modes of production and circulation of knowledge that change, highlighting in particular the role of relations between agents in the dynamics of knowledge creation, exploitation and distribution.

5.3. Building knowledge: economic intelligence

The transition from information to knowledge is the result of a collective process, focused on the common success of problem solving. It should be noted that this principle refers to the notion of collective intelligence (Dupuis 2004). The notion of knowledge calls for collective, community-based processes that are understood at the organizational and managerial levels and in relation to the company's strategy. According to Moinet (2009), "knowledge is a cumulative process if one has the capacity to exploit it". Similarly, Fayard (2008) proposes the notion of "strategic, action-oriented knowledge communities". Thus, EI is an approach inherent to the company's overall activity. Moinet offers management of the organization by EI (Economic Intelligence). In other words, there seems to be a need to pilot managerial and organizational practices through EI, in order to develop a learning organization, based on the creation and use of knowledge.

5.3.1. *The EI process and the transition from information to knowledge*

The company operates in an environment where information circulates rapidly and in large quantities. Research, processing and use of information have always been at the heart of the company's development. The information age puts the company in front of a double constraint: it must seek it and transmit it. Information is an input and an output and thus forms a process that according to Lesca (1982) is the process by which the company informs itself and its environment, and by which it informs its environment about itself.

The information collected and exploited allows the company to feed its reactivity and to initiate the changes that may remain. The search for information is based on two practices:

– the definition of a research plan;

– the mobilization of resources to collect the necessary data.

The exploitation of the collected data requires sorting, verification, processing, analysis and synthesis. It is through this process that the raw data collected during the research will be transformed into strategic information. The interpretation of the information collected through its processing, analysis and synthesis represents the lever for the success of the business intelligence process.

In her research in 2008, Lesca addressed the problem of interpreting data to transform it into strategic information, or knowledge. Interpretation systems, the heart of the EI process, are defined as the set of systems for attributing meaning to the information and stimuli that the company receives, manipulates and preserves (Baumard and Benvenuti 1998). The dissemination of information is crucial as it allows decision-makers to access the strategic information produced in and store this information for later use. The challenge is to give the right people the right information at the right time. This last notion implies that the information cycle should be optimized so that needs can be identified instantaneously, and the data collected can be processed as quickly as possible. This highlights the interaction between a multitude of actors (decision-makers, analysts, advisers, technicians, etc.) within the framework of a group dynamic, allowing the complementarity of each member's knowledge, for a better understanding, analysis of situations and production of the information necessary for action. Indeed, "the quality of the production of knowledge for operational purposes depends on the skills of interpretation and analysis of the human factor in collective problem-solving situations" (Bulinge 2006, 2007).

Business intelligence is then more than just mastering strategic information. It represents a process that contributes to changing the representations that the actors of the company have of their environment. In this sense, the strategic and collective dimension of EI lies in the triptych: appropriation, interpretation and action (Saïd 2004; Moinet 2009). Indeed, EI will enable the company to anticipate new growth levers and opportunities for the company, which will seize them by modifying its activities. Innovation is therefore facilitated here by an anticipatory vision of the business environment and an analysis of the room for maneuver available. And in order to make the company more innovative, the strategy must be adapted thanks to the mastery of decisive information. EI (Economic Intelligence) analyzes the information and allows us to anticipate evolutions and make decisions in conformity with the law.

It is also the perception of its environment, the identification of market developments, strategic adaptation, innovation and therefore sales growth, the opening of new markets and, the company's development strategy. Finally, it is, in a certain transversality, information security, data protection, risk anticipation and therefore the control of its information system which includes, today, the influence of the latter in and beyond its networks, also called e-reputation. All the fields that complement business intelligence, such as knowledge management, information protection and lobbying, can be grouped together in the overall concept of strategic intelligence. Strategic economic intelligence is a mode of governance whose purpose is the control and protection of strategic information relevant to any economic actor.

5.3.2. *Managing the data warehouse to extract knowledge and insight*

The wave of computerization that has swept over our societies is now reaching a form of maturity, and looking back on this flow, we cannot but be struck by the extent of the transformations that will have affected all sectors of information, creation and knowledge. Such a massive wave has developed very rapidly and has brought new uses to man, who has always needed to be informed. It is clear that information and communication technologies (ICTs) have participated in the evolution of this environment through the accumulation of large volumes of information and data (Big Data) from which knowledge is generated.

Big Data issues are related to volume of data, variety, processing speed and value. There are essentially three types of challenges around Big Data:

– the storage and management of massive data, on the order of 100 terabytes or petabytes, which exceed the common limits of traditional relational databases from the point of view of data storage and management;

– the management of unstructured data (which often constitutes the bulk of the data in Big Data scenarios), i.e. how to organize text, videos, images, etc.;

– the analysis of this massive data, both for reporting and advanced predictive modeling and for deployment.

According to Mark Breyer, co-author of Gartner's "Information Management in the 21st Century", "By 2015, organizations that have built a modern information management system will outperform their competitors financially by 20%". It remains to be said that without a doubt the processing of these masses of data plays a primordial role in tomorrow's society, as it finds applications in fields as varied as science, marketing, customer services, sustainable development, transport, health and education. Recently, there has been an emerging interest in the development of new analytical techniques specifically designed to address data mining and text mining issues.

The business intelligence process, with its different stages, is conditioned by a fluidity of information flows, a culture of group work, information sharing and knowledge exchange. But the question that arises is how to extract knowledge and insight from the large volumes of information that is at the heart of EI? The answer to this question requires even more mastery of the useful information that allows us to act in our economic environment, to ensure the protection of heritage and know-how, in order to gain a competitive advantage. Because of its growing use, the EI field is increasingly emerging as an important and rapidly growing area where important theoretical advances have been made. It becomes more and more useful as soon as we are faced with a large amount of information stored in the form of databases. Information, as a crucial resource for innovation, enables competitive intelligence, which is the result of the intelligent use of information and communication technologies (ICTs).

ICTs bring constraints to which companies must adapt a continuous flow of information, a much faster circulation of information, increasingly complex techniques. The risk is to be overwhelmed by this information, to no longer be able to distinguish the essential from the negligible. In fact, with the advent of the new market-dominated economy, the company's industrial problems have become more complex. In this perspective, economic intelligence is not reduced to the management of information flows, but is fundamentally interested in its interpretation, in its use to create knowledge. The ubiquity of the Big Data phenomenon further complicates the information literacy that can lead to the extraction of knowledge and insight. EI practices, on the other hand, make it easier for companies to measure, manage and increase the value of information, thereby guiding the decision-making process to achieve the best results through the cross-cutting and global sharing of knowledge.

This requires us to rethink concepts of the value of information in a world moving towards digital interoperability, sharing and collaboration. Transparency, participation and collaboration are the main issues in integrating the different economic actors in the open data paradigm. Today with the power of massive data analysis and calculation, it is possible to make digital information open and shareable for all, or what is called open data. Open data from both public and private sources adds a new dimension to data warehouse analysis and leads to new ways of innovation. The publication of open data joins the more general term open knowledge by the Open Knowledge Foundation, which defines it as knowledge, i.e. a work, in particular musical, cinematographic or data such as a scientific article, a geographical, governmental or administrative data. The term open is defined as the freedom to use, reuse and redistribute the original work freely.

5.4. Data mining, Statistica and Tibco[1]

The concept of data mining has gained popularity as a marketing management tool since it is expected to highlight known structures that can help decision-making under conditions of uncertainty. Data mining is often viewed as "a mixture of statistics, AI [artificial intelligence] and database mining" (Cortes and Pregibon 1997). Because of its increasing use, this field is increasingly becoming an important area (even in statistics), with strong growth, where important theoretical advances have been made.

But to explore these data, human skills, which include both the mastery of statistical techniques and analytical tools, are increasingly required in the business world: we speak of the "data scientist". But to have a chance to understand the data and take the first steps towards more effective treatments, data scientists must embrace the Big Data.

The term "data science" was coined by William Cleveland in a 2001 programmatic article entitled "*Data Science: An Action Plan for Expanding the Technical Areas of the Field of Statistics*". Data science is a new discipline that includes elements of mathematics, statistics, computer science and data visualization.

The latter or "data visualization" means the art of representing data visually. This can take the form of graphs, diagrams, maps, photos, etc. The presentation in an illustrated form makes the data more legible and understandable. Beyond illustration, data can even be presented in the form of animations or even videos.

The value of these data wave terms, which we understand, is not just "Buzzwords"[2] that attract attention, but they are also terms that give more value to data. We can now combine huge amounts of data to generate even deeper and more creative ideas.

---

1 O. LEBRET: Senior Consultant – Analytical Solutions @Statistica_fr – @tibco: #DataMining #BigData #Megadata #Scoring #DataScience #Analytics.

2 A Buzzword is a neologism generally designating a new word or a buzzword in a given universe of activity. See the site: https://www.definitions-marketing.com/definition/buzzword/.

But when we talk about Big Data, it does not just mean we are talking about the size of the data. The word "Big" also indicates the importance and the impact of the processing of these data because, with the opening of the data, the quantity will increase more and more and therefore "Big Data" says "Think Big"[3]. So, this universe that revolves around data now represents a movement in which France, with all its assets, must participate. But to meet this challenge, all the players must mobilize around this great movement in order to transform the risks linked to "Big Data" into opportunities.

5.5. Information an economic good?

Information is becoming a strategic issue not only for state security but also to defend a country's global competitiveness. This importance of information for the economic world has only recently become apparent, as we have just seen. Economic information is not a commodity like any other in the knowledge society. A company will be better than its competitors if it has, before others, the right information at the right time, whether it is market knowledge, legal, technological, normative or other information. To increase its competitive advantage, the company must be able to create an asymmetry of information to its advantage. Economic intelligence aims at a better mastery of information in order to be an instrument of decision support. Information, even if it appears derisory, can be of economic value after processing and cross-checking.

The information is therefore an economic asset for the company.

5.5.1. *Innovation as a driving force of growth*

The public resources (financial, human, etc.) mobilized in the framework of the support to innovative projects must be directed more particularly, towards the targeted phases of the process, and their effects must be able to be appreciated in return. These two imperatives combine to build the

3 The Think Big is a project initiated by Robert Muldoon when he was Prime Minister of New Zealand (1975–1984) and his New Zealand National Party government in the early 1980s as part of an interventionist policy (Wikipedia).

foundations of an effective public policy to support innovation. Thus, the complexity of the process and the various bodies (public and private) involved in it contribute to limiting the visibility of the public authorities' involvement, and above all that of the returns on public investment (in economic, social or financial terms for local authorities). Business intelligence puts into perspective all this information useful to an organization, whether it is scientific and technical or concerns markets, products, competitors, regulations and behaviors. The collection and processing of information, concerning the public resources mobilized, the identification of the privileged areas of intervention in the process, the assessment of the results and their impact on the local industrial fabric contribute to enlighten the public action by the construction of a certain number of indicators. In order to construct these indicators, it is necessary to set up relevant and exploitable databases and to implement the statistical tools to produce easily interpretable results.

Innovation, information and communication technologies, the service economy, use new engines of growth, which are information, knowledge, creativity and technologies. These new raw materials define what some call the economy of the immaterial.

5.5.2. *Strategic business intelligence*

Strategic economic intelligence is a mode of governance whose purpose is the control and protection of strategic information relevant to any economic actor. It can be:

– offensive when it collects, analyzes and disseminates information useful to the various economic actors; or

– defensive when it protects business-critical information from internal and external malicious and negligent acts.

> In order to prepare the new generations for EI, the Ministry of Higher Education and Research and MEDEF have signed an agreement to launch the skills repository "Economic Intelligence and New Risks of the 21st Century" Alain Juillet. The aim of this repository is to raise awareness and train students on issues of citizenship, economic intelligence and global security. This repository appears to be an important tool for awareness raising and integration for university and engineering school students.

The EI baseline and the new risks for the 21st Century.

It addresses the following key concepts:

– the value of information as fundamental data for the existence and performance of a company;

– the competitive and challenging environment of the company, presenting around some major issues of international competition, its opportunities and threats;

– the concept of risk to organizations and communities;

– the place and responsibility devolved to each citizen in the prevention and resolution of crises.

It has been designed to be deployed in university curricula as well as in engineering and management schools at the bachelor, master and doctorate levels.

Box 5.2. *Business intelligence and new risks for the 21st Century*[4]

The expansion in the number of data sources available, particularly through the Web, and the amount of data managed within these sources, has made it essential to develop systems capable of extracting knowledge, hidden by the complexity of the data. This complexity is mainly due to the diversity, dispersion and large volume of data.

But it will continue to grow at an exponential rate in the years to come. It therefore becomes necessary for the company to meet the challenges that allow it to position itself in the market, acquire market share and maintain a competitive advantage. The challenge is to be able to cross-reference as much data as possible in order to extract the best elements for decision-making.

The data exploited generate billions of dollars according to various reports (see the McKinsey Institute report). He recently indicated that

4 Born from a proposal of the 2003 Carayon report, the EI training repository was organized by the HRIE. The objective of this reference system is to identify the knowledge and know-how that a student should acquire in the context of a high-level, professional type of training in EI. The referential lists the major issues that the student must grasp and that a training course must provide. See: https://portail-ie.fr/resource/textes-de-reference/649/referentiel-de-formation-en-intelligence-economique.

opening "open data" to the public would allow the United States to save 230 milliards dollars by 2020 by allowing start-ups to develop innovative services in order to reduce only unnecessary energy expenses.

McKinsey & Company is a management consulting firm founded in Chicago in 1926 by James Oscar McKinsey, a renowned professor at the University of Chicago. McKinsey has established a Global Institute, located just steps away from the White House in Washington, D.C., to study macroeconomic trends. Some of his work is published, in particular through his journal McKinsey Quarterly.

Box 5.3. *McKinsey and Company*

The challenge with Big Data is to study these data and build models that can provide analyses that are relevant to the business issues. In turn, these data can be used as explanatory variables in "models" that are used by various users to make forecasts or, more precisely, correlations about the past that they can then project into the future.

If data are collected and stored very often, it is because they have a high commercial value for those who appropriate them. They enable them to target their offer more and more precisely to individuals based on this data. These are indeed the basis of "intelligence" in the most general sense of the word.

5.6. What is data science?

Data science is a multidisciplinary approach to finding, extracting and revealing patterns in data through a fusion of analysis, domain expertise and technology. Data science includes the fields of artificial intelligence, data mining, deep learning, forecasting, machine learning, optimization, predictive analysis, statistics and text analysis.

But what exactly is "data[5]"? Data are facts, observations or raw elements that are not organized or processed and do not convey any particular meaning. A data item represents an elementary and recorded element of a thing, event, activity or transaction. It is not significant in itself, but it only becomes significant when someone interprets it and gives it meaning.

5 According to TIBCO: O. LEBRET.

The way of "seeing" data and data analysis will gradually change at the dawn of the digital age. Thanks to the latest technologies (Smart, 4G, Cloud, etc.), the advent of the Internet, the various data processing and data mining software, etc. because the potential for exploiting this large amount of data and analyzing it has never been as promising and lucrative as it is today.

In order to stir up more and more data and move towards real-time analysis, it is first necessary to reveal the awareness of the rapid expansion of a new universe made up of data whose processing would be a real asset. Generally speaking, the technological evolutions linked to information processing capacities, throughout the information processing chain, lead us to wonder about the current interest in terms linked to "Data"[6].

One of the terms linked to the word data is indeed "Big Data", a phenomenon of interest to both companies and researchers, which brings constraints to which companies must adapt: a continuous flow of data, a much faster circulation of data, increasingly complex techniques. Big Data represents a challenge for companies that wish to develop their strategies and decision-making processes by analyzing and transforming this data into strategic information. In these new orders of magnitude, data capture, storage, retrieval, sharing, analysis and visualization must be redefined.

5.7. Conclusion

The considerable increase in the volume and diversity of digital data generated, coupled with Big Data analysis technologies, offers significant opportunities for value creation. This value is not only about what can be solved or improved, but also about what new discoveries will be made as a result of possible crossovers and correlations. To create value from these gigantic sources, we need algorithms, and to make these algorithms, we need

6 The "Buzzwords" around the Data family summarize "buzzwords" and data-related terms that people are unfamiliar with and whose use in conversation will seem to increase the "capital" of interest. In this context, "Big Data", "open data", "data visualisation", "data scientist", "data mining" and others are new terms that have replaced the word "data" or "data" as the focus of companies' concerns. These areas help drive the company's business through the "data mine" and represent paradigms for radical change that could lead the company to define offerings based on available data.

access to some exceptional talent. Thus, "we must create intelligence from the bytes that circulate"[7].

Huge amounts of data are transmitted over the Internet every day. We ourselves constantly produce them during each of our connections or transactions. These data have a value, and this value grows as processing, analysis and visualization tools are developed, which allow knowledge to be explored from the correct reading of these flows. Companies that are able to read the available data well benefit from relevant information and have the opportunity to innovate.

7 G. Recourcé, Scientific Director of EVERCONTACT.

6

From Knowledge to Strategic Business Intelligence: Decision-Making

"We have to change the company's culture!" The decision-making system is not a fixed system. It has to adapt and evolve over time, and to do this, it must:

– share information;

– evaluate the quality and relevance of decisions;

– challenge itself to reformulate, change and improve.

6.1. Data valuation mechanisms

Since the advent of computers and the Internet, the amount of data stored in digital form has been growing rapidly. More and more data silos are being created around the world. Individuals are making more and more data available to everyone via the Web. Many companies collect information about their customers and their behavior. Many industrial and commercial processes are nevertheless controlled by IT. Medical test results are also kept for analysis.

The increase in data produced by companies, individuals, scientists and public actors, coupled with the development of computer tools, offers new perspectives for analysis. Thus currently, not only is the amount of digitally stored data very large, but also the type of data is very varied. The data analyzed is no longer necessarily structured as in previous analyses, but can be text, images, multimedia content, digital traces, connected objects, etc. The data analyzed is not necessarily structured as in previous analyses.

In view of this volume and this diversification, it is essential to develop techniques to make the best use of all these stocks, in order to extract the maximum amount of knowledge from them. There are several techniques such as "data mining", which are not new, but follow the principles of descriptive and predictive methods. This is an integral part of data analysis, especially when the data is voluminous. Companies are increasingly aware of the potential contributions of Big Data in terms of developing new products or services, transforming their business model, improving customer knowledge or exploring new areas of development.

To get the most out of Big Data, the issue is not limited to the "simple" technical issues of collection, storage and processing speed. The exploitation of Big Data requires reconsidering the processes of data collection, processing and management. There is a need for governance to classify data and prioritize analytical priorities. All companies have a source of value through their data; Big Data is a generation of value as an analysis of this data.

It is the "analysis" applied to the data that will justify the Big Data, not the data collection itself. Companies must therefore be given access to the "information" that is likely to generate the most value. However, the process during which information acquires value is that of "Business Intelligence". It becomes increasingly useful as soon as the company has a large amount of information stored in a database. This is the best way for the whole company to take ownership of Big Data to create value. If data enhancement is emerging as a new strategic issue, governance is a requirement to achieve it.

6.2. How do you value data?

The amount of data (structured and unstructured) that comes from various sources and is produced in real-time leads us to say that there is a "Malthus law of data". Malthus had noted that the quantity of food increased only arithmetically, while the number of human beings increased in a geometric progression (Malthus 1798). Similarly, Pool (1984) noted that the supply of data increases exponentially while the amount of data consumed increases, at best, linearly.

This explosion of data volumes will grow as the "Internet of Things" develops. Human beings are relayed in their "data production" by a multitude of sensors in their homes and means of transport. At the same time, the production of data outside the traditional boundaries of the company (and their information systems) is growing exponentially, particularly via social media. The data produced is not very structured (photos, videos, etc.).

The data market is therefore in a situation where the quantity offered is far greater than the quantity requested. This is mainly due to the fact that our mental capacities and our time available to process information are constrained. The increasing digitization of our activities, the ever-increasing capacity to store digital data and the resulting accumulation of all kinds of information are generating a new line of business whose purpose is to analyze these large amounts of data.

Big Data has been described as "the next challenge for innovation, competitiveness and productivity" (Manyika *et al.* 2011). It offers companies unprecedented opportunities to increase revenues, performance and knowledge related to their business, market and customers. Social media data is a perfect example.

By leveraging technology to capture data from Facebook, LinkedIn, Twitter and others, the company can refine brand management, discover new distribution channels and strengthen customer relationships. The accumulation of data through various IT operations is a value potential that companies are not always aware of. Even if they do not necessarily know how to exploit them themselves, they sometimes hold resources that they do not yet value, even though these data and their use are a crucial issue for companies.

The challenge of data value is to specify and determine which data are intelligently usable. Solutions and methods have been implemented in companies such as Google, Amazon and others, where "Big Data" really makes sense. These players are characterized by their strong capacity for innovation and in particular their "data science". It is imperative to take stock of the mapping of the data held by the company and its ability to extract value from it. It is necessary to define a "priority business", implement the solutions to meet it and see if this system is labeled Big Data or not.

Exploitation is therefore the biggest issue, or, in another way and in the context of the Big Data phenomenon, the issue is knowledge and prevention through the cross-referencing of many data, and not a possible value generation. Companies need to be aware of the wealth of data they have available internally, and how to use it to derive value from it, before they want to add external data to it. It is no longer a question of identifying the data we need to store, but what can we do with that data? Capturing and cross-referencing this data becomes highly strategic to reveal new knowledge.

The analysis of large volumes of data can yield valuable insights that will help create competitive advantage. But this requires tools that process data in large volumes, at high speed and in a variety of formats. This calls for an efficient and cost-effective infrastructure: "cloud computing". The cloud is becoming a reality in business life. It provides a means that supports the volume of data and advanced analytic applications that can add value to the business. The development of cloud computing is closely linked to the Big Data phenomenon.

So, we need to know how to scale up these technologies to the scale of billions of units of information.

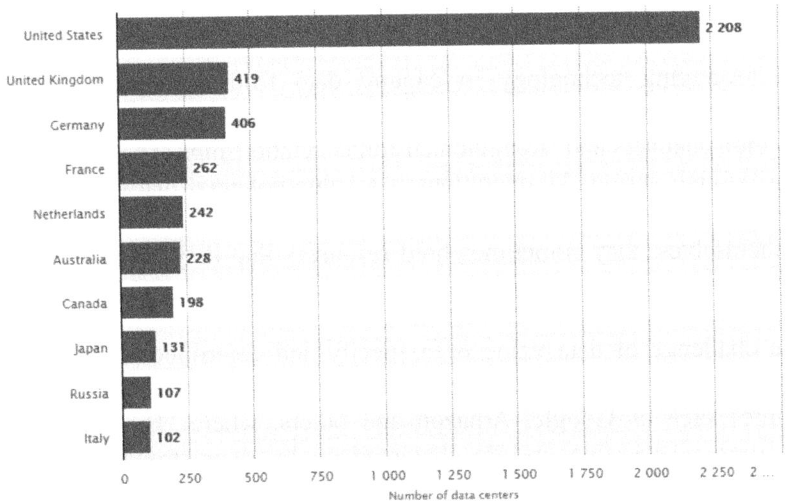

Figure 6.1. *Data Centers worldwide. For a color version of this figure, see www.iste.co.uk/monino/control.zip*

The major players, GAFAMs, are developing specific architectures in data centers to store large amounts of data at a lower cost. Data analysis models are able to provide more accurate results.

Box 6.1. *Digital data centers or factories and number of data centers worldwide by country in 2018 (source: Statista)*

It is also a question of developing a culture of data, its logic of exploitation and its visual interpretation. The challenge today is to develop technologies that enable massive data visualization flows. Visualization is a kind of use of the results of the analysis for their understanding and exploitation by operational decision-makers. Visual restitution allows the level of abstraction necessary to grasp the "mass data" and make sense of it. Correlations between the data make it possible to extract new knowledge.

The development of increasingly powerful and faster tools is pushing decision-makers to demand fast, effective and visualizable "decision aids" at any geographical point and therefore on the Web. In our information and communication society, statistics are becoming increasingly important. Statistics are omnipresent every day, especially through the media, each of us is exposed to a wealth of information from quantitative studies, opinion polls, "barometers" of popularity of politicians (Monino and Boya 2016), satisfaction surveys (Monino and Cavagna 1999), etc.

Moreover, for many functions, working life requires reading and understanding a wide variety of statistical data: market studies, dashboards, financial statistics, forecasts, statistical tests, indicators, etc. Statistics are the basis of the economic and social information system. Today, it is extremely accessible, thanks to the tremendous development of databases, particularly via the Internet, or easily constructed using spreadsheets integrated into the most common software. Among the techniques available are the following:

– Descriptive techniques: they aim to highlight information that is present but hidden by the volume of data (customer segmentation and product association search on sales receipts). Thus, these techniques make it possible to reduce, summarize and synthesize data, such as:

- factor analysis: projection of the data in graphical form to obtain a visualization of all the links (reconciliations and oppositions) between the different data;

- automatic classification (clustering – segmentation): association of homogeneous groups of data allowing to highlight a segmentation of individuals in classes;

- search for associations (analysis of the sales receipt): this involves detecting dependencies between the objects or individuals observed.

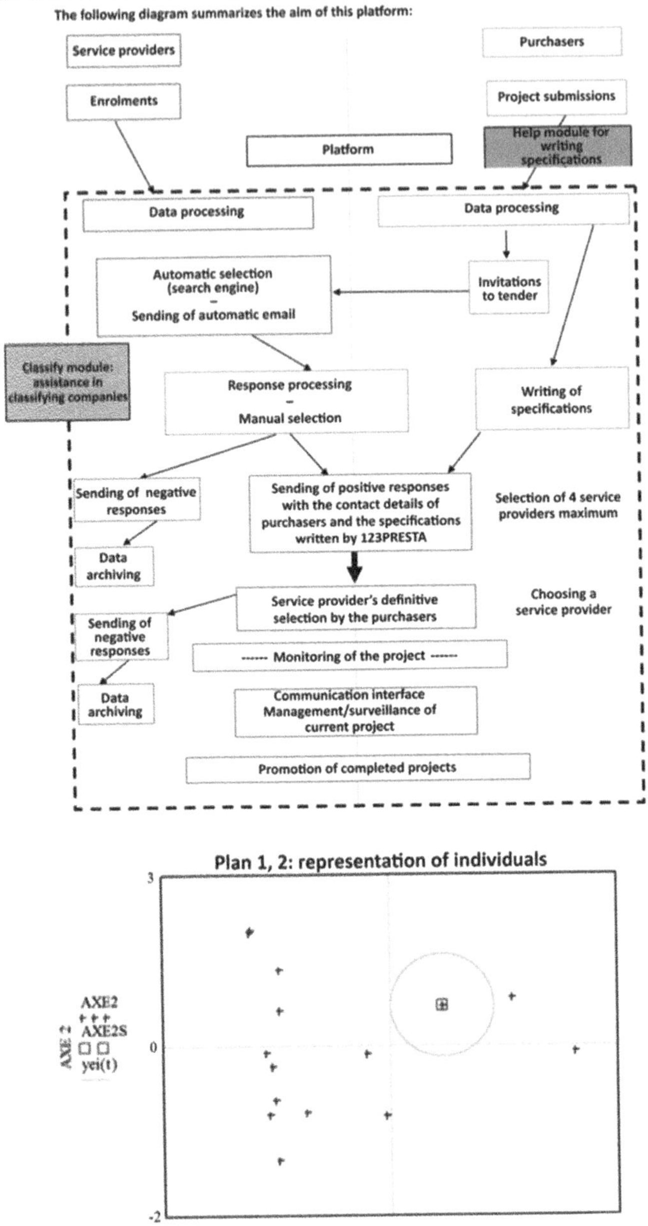

Figure 6.2. *A classification by data analysis concerning a group. For a color version of this figure, see www.iste.co.uk/monino/control.zip*

It is important to emphasize in this conclusion that today there are barely any specialized directories that are this qualified and up-to-date. The objective remains the same: to become the leader of intermediation in France and to publish the figures of an ICT observatory at the national level worthy of the name. The forecasts for the next two years are quite modest in view of the potential of the national and then international market. There are now just over 10,000 service providers that can be listed in our directory.

Research and development work carried out by the TRIS laboratory and supported by Transfert-LR.

Box 6.2. *An example of data processing for the start-up 123PRESTA in 2010*

– Predictive techniques are analyses that enable realistic projections to be made in order to identify new sources of opportunities (or threats) and anticipate responses adapted to the real situation. They are intended to extrapolate new information from existing information. These statistical methods look for a relationship between one or more dependent variables in relation to independent variables, qualitative (discriminant analysis/logistic regression).

– Decision trees, make it possible to divide the individuals in a population into classes; the first step is to choose the variable that best separates the individuals in each class into sub-populations called nodes, depending on the target variable.

– Neural networks or machine learning from the field of artificial intelligence is a set of nodes connected to each other and possessing values. Learning is done by adjusting the weight of each node until an optimal solution is found to reach the set number of iterations.

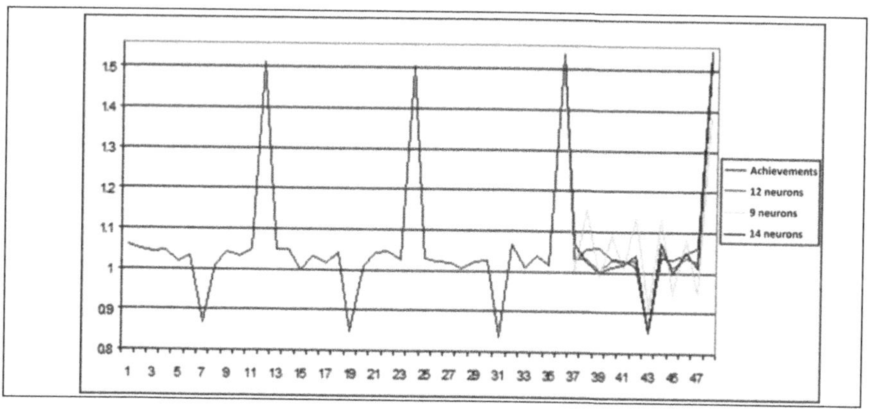

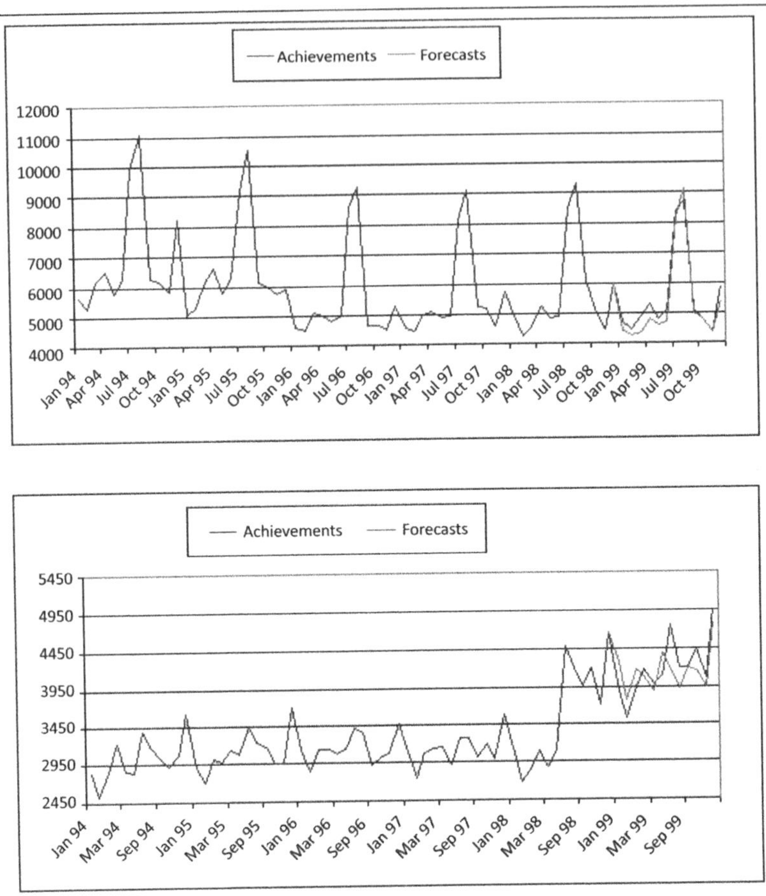

Figure 6.3. *Predictions using the neuron method. For a color version of this figure, see www.iste.co.uk/monino/control.zip*

Within the framework of an exchange under agreement between the laboratory and the European distribution giant, two methods for mass forecasting were tested:

– the neural method “Machine Learning”;

– the classic forecasting method.

Box 6.3. *Time series forecasting with neural networks. Turnover of mass retail stores*

– Simple or multiple linear regression allows us to model the variation of a dependent variable with respect to one or more independent variables and thus to predict the evolution of the first one according to the latter.

These techniques will be developed and enriched to improve data processing. Thus, the software required for this processing must be able to detect interesting information: the ability to combine more data will strengthen the cross-functionality in the company, which increases reactivity. Another aspect of the "Big Data" revolution, which reinforces the power of mathematical formulas, is that it allows data to speak for itself; we are talking about "algorithms" here. Beyond data collection and storage, algorithmic intelligence is essential to make sense of the mass of data. Algorithms allow correlative models to be established by ordering, sorting, prioritizing data, making them intelligible via a correlation or prediction model.

Companies can now understand complex phenomena and, above all, share these analyses to increase their collective intelligence. However, it should be noted that each company must define its own way of using the data. To do this, it needs a factor that is often referred to as a "data scientist". He or she must be able to analyze huge amounts of data and find correlations. However, these correlations must be applicable, cost-effective and achievable. The necessary qualities are therefore data skills, creativity in interpreting data and validating correlations, knowledge of the company in order to master the use or find models that improve profitability[1].

The "Big Data" phenomenon thus raises many questions about the technological development it will bring about and the value it is likely to create. It requires the development of new multidisciplinary training programs, leading to operational know-how particularly adapted to the "Big Data" field, at the interface of computing, statistics and mathematics, as well as a strategic overall vision for the design of new services or products and the deployment of advanced decision-making information systems.

---

1 A statistician has become a data scientist or data analyst: whenever the volume to be processed is enormous, the analyses must be carried out with a view to improving the efficiency and profitability of the company. The emergence of "Data Science" in companies and the understanding of corporate data is a strategic asset that can become a powerful vector of differentiation and performance. In the long run, it is a question of making sense of massive data deposits and highlighting their added value.

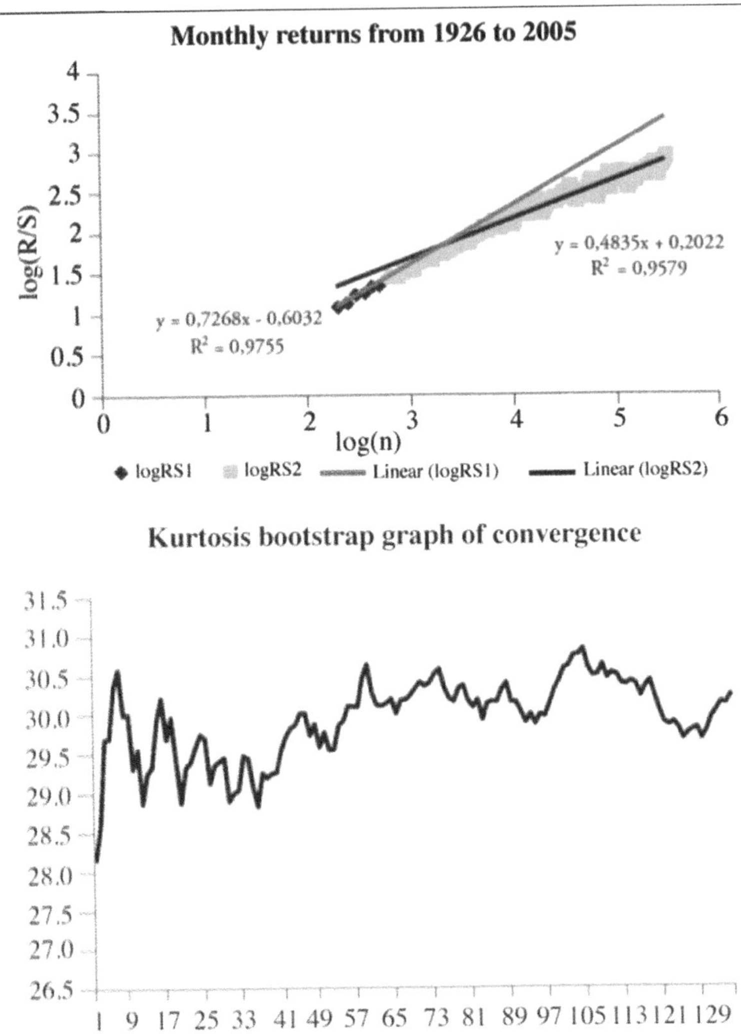

Figure 6.4. *The stock market, chance, deterministic chaos*

Since 1985, "deterministic chaos" has entered stock market studies. Various studies have attempted to test its relevance to random walk, which, it should be recalled, underpins both FAMA's low efficiency theory and the various "models" claiming to represent the functioning of markets.

We have used French long-term data (1965–2002) and we have tried to purify daily productivity, using the Hurst exponent calculation, over 13,000 points.

Box 6.4. *Chaos, Hurst and bootstrap exhibitors. An example applied to the Paris Stock Exchange (Matouk and Monino 2005)*

Another important point in the data valuation process is the "quality" of the data. These data, which are the basic elements of the processes, constitute an essential asset and asset for the company. The quality of the information resulting from a Big Data processing approach is directly linked to the quality of the datasets entered, whether they come from inside or outside the company. Data quality is an interesting key in the context of the Big Data phenomenon because based on unreliable data, bad decisions will generate bad choices.

"Data" are no longer structured and relational formats, they are referenced as unstructured and heterogeneous content (comments, videos, images, sounds, sensor data, etc.). On the other hand, image or object recognition, defining explanatory variables, resolving semantic subtleties and cross-referencing different data sources are unstructured analysis.

Thus, "open data" is another source of data; it is not another quantity that is added to the Big Data, but rather reliable and reference data. Volume is not a real issue: the challenge of Big Data is to identify intelligent data and to be able to interpret it to improve the performance and competitiveness of the company. The challenge of interoperability is to be able to cross-reference a very large amount of data from fragmented databases. As such, an "Open data" approach represents a new business model in which the company becomes a data producer.

In a context of crisis and accelerating cycles, companies must constantly optimize their productivity and operational efficiency. Poor data quality (customer, product, supplier, structure, etc.) will have a direct impact on the company's competitiveness, efficiency and responsiveness. To transform data into value, it is necessary to invest in technology: mastering techniques, methods and tools will undoubtedly fulfill this function. But technology is not enough to solve the problem.

Collecting data without a prior logic, without defining a specific strategy, can be much less profitable than expected. The absence of unified and controlled management can also have serious consequences on the lack of control of operational risks. This requires a real governance policy that goes beyond collection and processing operations. The aim is to help decision-making in a context where "information", which is at the heart of "economic intelligence", has become the major strategic asset of companies.

The exploitation of the masses of data collected requires sorting, verification, processing, analysis and synthesis. It is through this process that the raw data collected during the research will be transformed into "strategic information". The interpretation of the collected data represents, through their processing, analysis and synthesis, the lever for the success of the "economic intelligence" process.

The process of data enhancement, the processing of which generates information with high added value for the company and aims to help companies take advantage of the "Big Data", consists of collecting, processing and managing massive volumes of data in order to clearly identify what they can bring to the company, as well as measuring and monitoring their use, user consumption patterns, storage and governance. The latter covers the previous steps and establishes processes and good operating practices as required.

6.3. Data governance: a key factor in valuation

It is difficult for companies to exploit the ever-increasing volume of various data. Taking advantage of the Big Data phenomenon, which can often be summed up in four words – the "4 V" or, to go further, the "7 V" – appears to be a major challenge. In addition, automatically generated data often requires new analysis techniques, which presents an additional challenge. According to available statistics, 80% of business data is unstructured. In addition, plain text documents, videos, images, etc. are added to the data types.

Text data, log data, mobile data, videos, etc. have disrupted traditional processing technologies because they are not structured data. The new challenge lies in the handling of these new data formats. Some of these resources are stored in databases, developed with a view to managing, retrieving and exploiting them. Once again, there is the problem of data quality.

In this respect, how can a company know that the data at its disposal are reliable and suitable for use, in order to be able to extract knowledge (see the knowledge pyramid?). This means that treating erroneous and unreliable data in the same way as other data is likely to distort the analyses and, of course, the decisions that follow. This data quality risk is at the origin of a new concept called "Smart data". Great importance must be placed on finding the right data (in the warehouse) that is reliable and useful, and that will provide information and create value, rather than focusing only on how it is processed.

In these circumstances, the company must seek a means of facilitating the processing of this volume of data and making it available for a wide range of analytical purposes. In addition, the in-depth discussions concerning the control of the company's information assets have led the company to adopt data governance principles. For an organization to move towards valuing its data as a corporate asset, it needs to change its culture and the way data is managed.

This means mixing the company data, already available and structured, with the Big Data, "unstructured". This is the first step towards data management or "data governance". Data governance is a set of functional areas that manage, integrate, analyze and govern data.

Data governance is defined as the ability to blend reliable and timely data to feed large-scale analytics, regardless of the data source, environment or user role[2]. It is a quality control framework for assessing, managing, operating, optimizing, controlling, maintaining and protecting business data. Based on this founding concept, users will be able to easily explore and analyze mixtures of validated data from across the enterprise.

Once the relevant data has been obtained, it must be properly formatted for analysis. Data ready for analysis must comply with all relevant governance rules so that it is reliable, compliant, up-to-date and properly blended. The introduction of analysis tools based on massive volumes generates:

2 Pentaho Business Analytics, Designing an optimized data refinery, An enterprise solution to provide mixed, prioritized and analysis-ready data on demand.

– access to existing data, but for uses not necessarily listed. These uses may lead to violations of existing security policies, particularly with regard to regulatory constraints (protection of privacy, health data, banking data, etc.);

– import from social networks, the Web … of new data not yet listed in the company. These new data call for a review of policies to take into account their particular sensitivity;

– the results obtained from the analysis are in themselves new data to be protected.

Security and governance are becoming critically important. In March 2014, Gartner published a report entitled "Big Data Needs a Data-Centric Security Focus". The group points out that: "Information security managers should not manage Big Data security in isolation, but need rules that encompass all data silos to prevent this management from turning into chaos". It recommends evaluating implementations of data protection solutions against security policies for databases, unstructured data, cloud storage and Big Data.

> The multiplicity and disparity of sources creates challenges for the collection, processing and storage of data. There are many uses: from marketing targeting (scoring[3], rebound, behavioral analysis) to a more technical use (log analysis), including activity management, particularly during the transformation phase. Google, Amazon and Facebook have demonstrated the power of data to improve customer relationships, sell more and differentiate themselves from competitors.

Box 6.5. *The challenges of multiple sources*

The introduction of a data analysis solution requires both a review of existing policies to incorporate new uses of the data and an extension of existing policies to incorporate issues specific to the new data. To do this, the life cycle of the data collected must be optimized so that the needs can be identified instantly and processed as quickly as possible. This highlights the interaction between a multitude of actors (decision-makers, analysts, advisers, technicians, etc.) within the framework of a group dynamic allowing for complementary knowledge: better understanding, better analysis of situations and production of information necessary for action.

3 Scoring et Statistica: "Le Data Mining en 35 Leçons avec STATISTICA" Tutorials https://www.youtube.com/watch?v=VDejgEK0FqA.

In order to ensure relevant data collection, it is essential to define for which analytical needs these data will be sought and collected, and before moving on to how to analyze these data, what techniques and tools will be used. It is a question of defining a transversal responsibility within the company itself. The transition from information to knowledge is the result of a collective process, focused on the common success of problem solving. It should be noted that this principle refers to the notion of collective intelligence (Dupuis 2004). The notion of knowledge calls for collective, community-based processes that are understood at the organizational and managerial level and in relation to the company's strategy.

With the huge volume of data that accumulates in most organizations, it is not always easy to find the right data or even to transform it to extract a useful analytical view. Effective data governance is essential to preserve data quality, as the volume and complexity of data grows as there are interactions within the company (the internal environment) and between the company and its external environment (customers, suppliers, competitors, etc.).

The first step in building a governance framework is to identify the extent of the need by conducting a gap measurement between existing practices and the defined target of good governance. This measure is structured according to the action levers available within the information system. In concrete terms, governance must provide the commitments, structures and resources, as well as ensure the methods and tools to be able to:

– manage all data and associated metadata;

– manage data consumption processes;

– identify data and design associated architectures;

– manage the data life cycle from a cross-domain view of the enterprise;

– improve data quality.

Data governance combines a set of people, processes and technologies to ensure the quality and value of an organization's data. The technological part of data governance combines data quality, integration and management. The success of a data governance strategy presents various technical challenges. First of all, companies need to integrate the different data sources, implement measures of their quality to improve functionality. They must then establish and improve the quality of their data. Finally, collaborative work is necessary for different teams to work together on data quality issues.

In other words, it seems necessary to pilot managerial and organizational practices in order to develop a learning organization, based on the creation and use of knowledge. This brings us back to a very important concept in the business world, namely "Business Intelligence" as a mode of data governance, which goes beyond a simple mastery of strategic information. EI represents a process that contributes to changing the representations that the company's actors have of their environment. In this sense, the strategic and collective dimension of EI lies in the triptych: appropriation, interpretation and action (Saïd 2004; Moinet 2009).

EI will indeed enable the company to anticipate new growth levers and opportunities for the company, which will seize them by modifying its activities. In order to make the company more innovative, the strategy must be adapted through the mastery of decisive information. EI analyzes information, anticipates changes, and enables decisions to be made in compliance. This is how information security, data protection, risk anticipation and the control of its information system, which includes its influence within and beyond its networks, is conceived. All the fields that complement business intelligence, such as knowledge management, information protection, lobbying, can be grouped together in the overall concept of strategic intelligence.

Strategic economic intelligence is a mode of governance whose purpose is the control and protection of strategic and relevant information for any economic actor. It can be:

– offensive when it collects, analyzes and disseminates information useful to the various economic actors; or

– defensive when protecting business-critical information from internal and external malicious and negligent acts.

With the advent of the new knowledge-based economy, the industrial problems of companies have become more complex. In this perspective, economic intelligence is not reduced to the management of data flows but is fundamentally concerned with its interpretation and use to create knowledge. The ubiquity of the Big Data phenomenon makes information literacy more complex and can lead to the extraction of knowledge and learning. EI practices, on the other hand, make it easier for companies to measure, manage and increase the value of data, thereby guiding the decision-making process to achieve the best results through the cross-cutting and global sharing of knowledge.

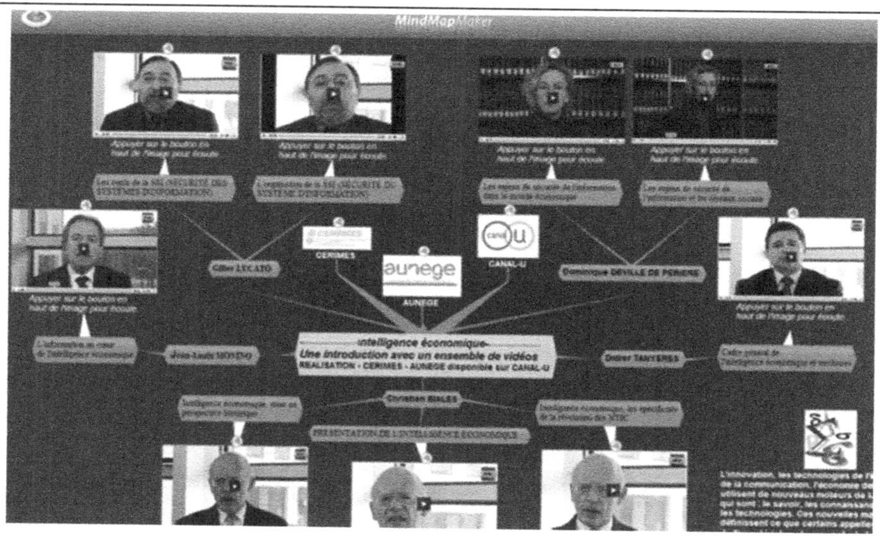

Figure 6.5. *Example of educational resources put online by AUNEGE. For a color version of this figure, see www.iste.co.uk/monino/control.zip*

Presentation of the concept of economic intelligence in nine educational videos. These educational resources are put online by AUNEGE[4] and available on Canal-U[5].

Box 6.6. *EI short presentation videos*

4 AUNEGE: Université Numérique Thématique en Economie et Gestion. http://www.aunege.org/.
5 Canal-U: La vidéothèque numérique de l'enseignement supérieur. Une web tv pour les enseignants et les étudiants. https://www.canal-u.tv/.

Our era is characterized by an exponential increase in data production and processing, as well as by a massive increase in data transmission speed and in the speed of access to stored data. We probably can't fight Big Data. But if we want to make a real profit from it, it is essential to control it by implementing a strategy that allows the data to be valorized. It is only with reliable data governance that corporate data can become a strategic asset, which ultimately translates into real competitive advantages and increased corporate value.

The main objective of data governance is to ensure that the operation of Big Data follows a formal strategy to achieve specific results: it ensures that all the right questions are asked in order to take advantage of Big Data. The governance of gigantic data therefore remains to be built within companies. This requires data prioritization and data protection, which in turn requires widespread adherence at the enterprise level.

6.4. EI: protection and enhancement of digital heritage

Today, the success of a company is often conditioned by the control we have over information. Access to information, its analysis, dissemination and presentation are elements that determine success. The information is now available very quickly and now within reach of all stakeholders. It becomes an economic heritage and represents a product of values generated by sharing and exchange. Thus, depending on its value, information can become sensitive or even strategic; it is therefore linked to the notion of secrecy because it represents a key factor affecting the entire economy.

This exponential increase in the amount of data available provides the opportunity to create new uses for this information. The value of all these data is nil if there is no upstream process for transforming the available information into analysis. In summary, the forces that accompanied the emergence of Big Data and that continue to develop it are[6]:

– automation of data exchange;

– a storage revolution;

– the advent of a new data science;

6 Big Data: "définition, enjeux et études de cas Data-Business.fr" http://www.data-business.fr/big-data-definition-enjeux-etudes-cas/.

– progress in data visualization;

– new opportunities for monetization.

These applications will be reinforced in the coming years, provided that companies adapt their operating methods to this evolution. To do this, they must constantly innovate, exploiting and cross-referencing the quantities of data from the various connected objects. It is then a matter of transforming the mass of data (available in different forms, often unorganized and collected through different channels) into information and then into knowledge. But currently, the information we are looking for is hidden within a gigantic mass of data.

The exponential growth of data, coupled with the use of algorithms, has contributed to the emergence of the concept of "Big Data". EI practices are not in a position to neglect this phenomenon, nor to ignore the revolution affecting their treatment. The success of an EI strategy, therefore, depends on the ability of companies to master and use the mountains of data they hold, or "data mining". This exploitation gives even more value to the information, which must be protected.

The use of Big Data allows us to cross data in order to obtain a precise cartography, which must provide quality information, allowing a company to evaluate the alternatives, to better design its way of acting and to ensure a secure management of its assets, within the framework of an approach based on collective intelligence. Companies now have to deal in real-time with a significant increase in the amount of data available that can influence the decision-making process.

A number of events help to situate the concept of EI[7] and thus shed light on its origins and evolution, identifying the conditions that presided over its emergence and development. The methods and tools of Business Intelligence allow the validation of the collected data (from different sources considered reliable) into a coherent set of information adapted to the profile of the company and its needs.

7 Business intelligence is not an innovative practice born exclusively out of the development of technology and the rise of the information society. Each country, depending on its history, geographical location and wealth, has led a process that can be described as economic intelligence. A remarkable witness of his time, "Sun Tzu" once said: "He who knows his enemy as he knows himself, in a hundred battles will not be defeated. He who knows himself but does not know the enemy will be victorious one time out of two. What about those who don't know themselves better than their enemies?"

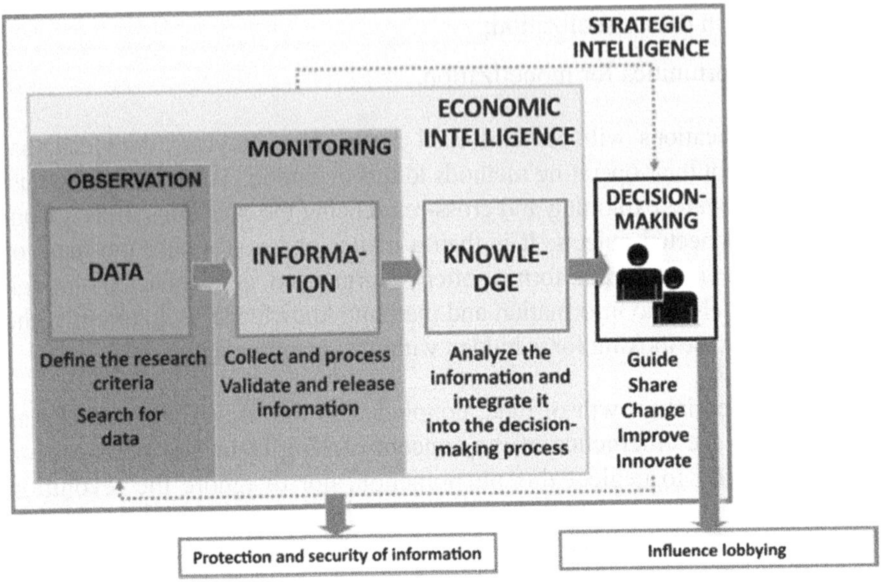

Figure 6.6. *A model of strategic business intelligence, influence and information security. For a color version of this figure, see www.iste.co.uk/monino/control.zip*

To manage such a mass of data and information, it is absolutely essential to adopt pragmatic and efficient sorting and selection methods. Economic intelligence covers a range of approaches from monitoring to the search for strategic information. These approaches are implemented by companies to monitor their environment, increase their competitiveness and manage the risks associated with their business. The advent of the Internet has, of course, facilitated access to a set of data that makes it easier to get to know the company's partners, suppliers and competitors.

As an example, a distribution group could use geolocation data from a mobile phone operator to identify the most promising areas. We will be there at the crossroads of business intelligence and marketing. Alternatively, a company active in the B to C sector could analyze large masses of comments produced by Internet users on social networks, in relation to its brand. This time, we are at the crossroads of economic intelligence and communication (Monino and Sedkaoui 2013a).

Box 6.7. *The place of geolocation data*

The EI process will allow an organization to be both reactive in adapting to change and proactive in taking a forward-looking attitude to better understand the dynamics of its environment. The implementation of an EI approach within a company within the framework of Big Data is a response to the challenges of an increasingly complex and unpredictable globalized environment. In this context, it is necessary to control the information flows of the entire company before taking a strategic decision.

The development of an EI approach within a company can only be envisaged with the participation of everyone. The confrontation of the different levels of responsibility, informed by a strategic line, is the best way to help the decision-maker make the most relevant choices at the right time. Assessing the quality and reliability of information and determining its usefulness to the business is probably the most important part of the business intelligence process.

To increase its competitive advantage, the company must be able to create an asymmetry of information to its advantage. Information is becoming a strategic issue, not only for the security of the state but also to defend a country's global competitiveness. The importance of information as a raw material for the economic world has become apparent only recently: economic information is not a commodity like any other in the knowledge society.

EI aims at a better mastery of information in order to be a tool for decision support. Even seemingly trivial information can be of economic value after processing and cross-checking. Information is therefore an economic asset or even a patrimonial asset for any company. In addition, from the point of view of the quantity of data, EI provides a global vision of the company's environment and monitors the entire information sphere (markets, technologies, etc.) and makes it possible to capture strategic information in the available mass. It has become one of the fastest-growing cross-cutting activities around the world and can be seen as both of the following:

– An informational approach comprising a set of operations or processes by which collected information becomes usable and worthwhile. This approach also aims to harmonize the processes of research, processing, distribution and protection, in line with the requirements of the context and the actors involved. The management of these actions is the purpose of the monitoring activity, which is essential to the economic intelligence approach.

– Mediation between the different actors involved:

- the watchdogs are responsible for providing, according to an explicit or implicit request for information, useful information in the best conditions of quality, timeframe and cost;

- decision-makers are involved in this process at different levels, and are most often solicited through several roles and functions.

EI can be seen as a new managerial practice at the service of the company's strategy, enabling it to improve its competitiveness. It is a response to the strategic questioning and concerns of decision-makers to solve their decision-making problems and improve the strategic position of the company in its environment. "To provide decision-makers with the necessary elements to better define their medium- and long-term orientations" (Salmon and Linarès 1997). EI produces an "Info Action" allowing for proactivity and interactivity (Bulinge 2002). This principle of dynamic use of information operates within a coherent structure and is based on a phenomenon of collective intelligence.

According to the report (Martre 1994)[8]: Economic intelligence can be defined as all the coordinated actions of research, processing and distribution, with a view to its exploitation, of information useful to economic actors. These various actions are carried out legally with all the necessary guarantees of protection to preserve the company's assets, in the best conditions of time and cost. Useful information is that which is needed by the various levels of decision-making in the company or community to

8 In France, many political figures have taken an interest in this notion and without a doubt Edith Cresson is the first to become aware of France's disarmament of this practice. Edith Cresson was the first in France to take an interest in EI when she was the Minister of Foreign Trade during these trips to Japan and the USA in 1988. It was in the 1980s, when she was the Minister of Foreign Trade, that Edith Cresson became aware of France's disarmament in the field of economic intelligence. In 1991, she took part in setting up the working group of the French Planning Commission (Commissariat Général du Plan), which led three years later to the publication of the Martre report: "Economic intelligence and company strategies". In 1998, Edith Cresson published "Innovate or Suffer", a book in which the European Commissioner for Research and Technology calls on France to innovate at all costs and in all fields. She "created the beginning of a culture in France". The latter was indeed at the origin of the impetus given to economic intelligence at the Harbulot national level.

develop and implement, in a coherent manner, the strategy and tactics necessary to achieve the objectives defined by the company in order to improve its position in its competitive environment.

These actions, within the company, are organized around an uninterrupted cycle, generating a shared vision of the company's objectives. EI is the result of a strategic intention and makes all the economic actors of a company, a community, a territory, national, transnational, States interact. This definition provides a global vision of the action environments in which EI (Monino and Boussetta 2013) must emerge. In particular, the coordination of these actions provides the company with visibility on its various relationships with its environment by constantly observing the behavior of its competitors and the realities of the markets.

The analysis of more data in real-time is likely to improve and accelerate decision-making in a wide range of sectors, from finance to health and research. The new analytical power is seen as an opportunity to invent and research new methods capable of detecting correlations within the amounts of data available. Anything related to data or business intelligence can be re-characterized as "Big Data", and it looks like this craze will soon reach the level of excitement generated by cloud computing, which has seen existing offerings rebranded and entire organizations moving to the cloud overnight.

6.5. Data analysis techniques: data mining/text mining

Today, we have much more data to process and manage: commercial transactions, scientific data, satellite images, text documents, text reports, multimedia channels, etc. Because of this voluminous increase in data, the design and implementation of effective tools that allow the user to access only the "information" that he or she considers "relevant" becomes an absolute necessity, especially since this information allows the user to better understand his or her environment, to be on the lookout for the latest market trends and to better orient his or her strategy.

The diversity and increasing volumes of data that are flowing into companies at an accelerated pace make it very difficult to access and exploit. Companies have a large amount of data at their disposal. These reservoirs of knowledge need to be explored in order to understand their meaning, to detect relationships between data and to have models explaining their behavior. Faced with huge data collections, we have created new needs that allow us to make the best managerial choices. These requirements represent the automatic summary of the extraction of information from the stored data in order to discover new knowledge.

To meet these needs, a set of architectures, approaches and tools, some new and some long-standing, has been grouped together under the term "data mining". The birth of "data mining", a term initially used by statisticians, is essentially due to the combination of the following two factors:

– the exponential growth in companies of data related to their activity (data on customers, stocks, manufacturing, accounting, etc.) and which contain vital information on their strategic functioning for decision-making;

– the very rapid progress of hardware and software.

The aim of data mining[9] is therefore to enhance the value of the data contained in companies' information systems. Exploiting a multitude of data transiting within their environment in order to make better decisions is often perceived as a key issue for companies.

"Data mining" is the discovery of new structures in large datasets, involving statistical methods, artificial intelligence and also database management. It is therefore a field at the crossroads of statistics and information technology whose goal is to discover structures in large datasets. "Data mining" as a process (as automated as possible) that goes from the elementary data available in a "Data Warehouse" to the decision, providing at each stage of this process an "informational added value" that can go as far as automatically triggering actions based on the summary information.

It is a process of automatically extracting predictive information from large databases. According to the Gartner Group, this process can be iterative and/or interactive depending on the objectives to be achieved. It can be said that the main task of "data mining" is to use methods to automatically extract useful information from these data and make it available to decision-makers. As a process, it is relevant to underline

9 In this case, "data mining" is positioned as the tool capable of carrying out this mission. Another term is "knowledge discovery from data" (KDD).

here that "data mining" does not only refer to highly developed tools and computer technology. It is indeed necessary to note the fundamental role of the human being, whose involvement must be total in every phase of the process.

"Data mining" is a technology for enhancing data and extracting information and knowledge. The growth of this technology is the result of an increase in digital data, which, due to its abundance, is under-exploited without adequate tools and expertise. This technology is based on a variety of techniques (artificial intelligence, statistics, information theory, databases, etc.) which require a variety of high-level skills.

The success of this discipline has grown, and interest in it has grown along with the size of the databases. Today, with the rise of the "Big Data" phenomenon, we can consider that we have entered a phase of mastery of this discipline. The challenge for companies faced with "Big Data" is therefore not so much in the capacity for analysis itself as in two issues that tend to be too often ignored[10]:

– the mode of data collection must remain known and controlled to ensure that a data mining analysis does not produce counterproductive effects for the company;

– the analysis of large amounts of data should not be done at the expense of their quality. Not all of them have the same utility and do not allow the company to generate value.

The types of models that can be discovered depend on the data mining tasks used. There are two types of data mining tasks:

– descriptive tasks that describe the general properties of existing data;

– predictive tasks to make predictions on available data.

But the heterogeneity of data sources and their characteristics means that data mining alone is no longer sufficient. With the evolution of the Web (the transition to the semantic web), we are witnessing an explosion of textual data, unstructured data that raises many potentialities for companies that cannot ignore its existence and impacts on their ecosystem (suppliers, competitors, customers, etc.). The Web is another reason for

---

10 http://www.decideo.fr/Le-Big-Data-un-cadeau-empoisonne-pour-les-Data-Miners_a5457.html.

the interest in extracting knowledge from texts. Indeed, with the Web, unstructured data (such as text) has become the predominant type of online data.

In this framework, useful information is not only to be found in quantitative numerical data, but also in texts. The tool for accessing and collecting textual data must be capable of operating from the Web on both HTML documents and databases, either bibliographic or textual. This is text analysis or "text mining" (Monino and Boya 2013), which is a technique for extracting knowledge from unstructured or loosely structured documents or texts, using different computer algorithms.

This means modeling linguistic theories using computer tools and statistical techniques. The use of computers to automate the synthesis of texts is not really recent. Hans Peter Luhn, a researcher at IBM, the true inventor of the term Business Intelligence in 1958, published a study in 1957 entitled "The Automatic Creation of Literature Abstracts". This exciting study can be accessed directly on the IBM research site.

Text mining consists of first recognizing words or sentences, and identifying their meaning and relationships, then processing, interpreting and selecting a text. There are two types of selection criteria:

– the novelty is to discover the relationships, including the implications that were not explicit;

– the similarity or contradiction in relation to another text or in the answer to a specific question consists of discovering texts that correspond most closely to a set of descriptors in the initial query.

"Text mining" (Monino and Boya 2010, 2011) then generates information about the content of a specific document; this information will be added to the document, thus enriching it. "Text mining" finds its main applications in survey processing or data analysis projects where some responses are in an unstructured or textual form (e.g. email messages, comments, suggestions in a satisfaction questionnaire with open-ended questions, description of medical symptoms by patients or practitioners, complaints, etc.) which it is desirable to integrate in the course of the overall analysis.

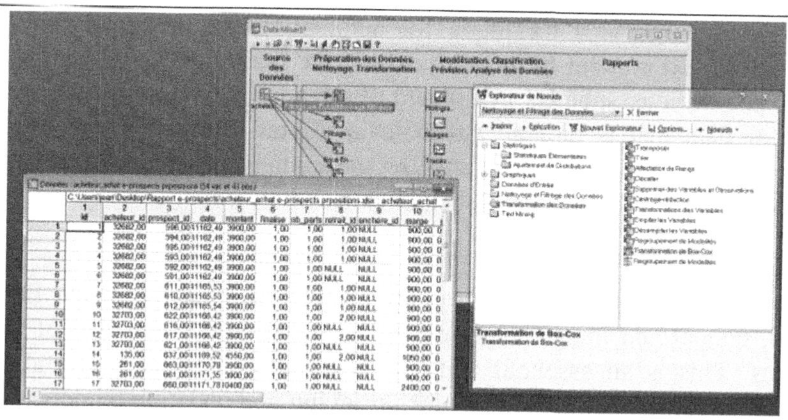

Figure 6.7. *Example screen of TIBCO's Statistica. For a color version of this figure, see www.iste.co.uk/monino/control.zip*

In the background, the data mining process of Statistica, on the left in the foreground the database and on the right in the foreground the different methods that can be used in the data mining window.

Box 6.8. *An example of data processing a database for a start-up with TIBCO's Statistica*

These techniques are frequently used to produce predictive models to automatically classify text, for example, to automatically forward emails to the most appropriate recipient, or to distinguish between "spam" and important messages. Text mining therefore makes it possible to:

– automatically classify documents;

– prepare a document preview without reading it;

– feed databases;

– ensure the monitoring of documents.

Two approaches can be considered:

– the statistical approach, which produces information on the number of occurrences of a word and the frequency of its appearance in a document;

– the semantic approach, which is based on an external element that is the repository, which can be static (keywords, etc.) or dynamic; it implements logics (information that is deduced from the repository).

"Text mining" is a technique for automating the processing of large volumes of text content in order to extract the main trends and statistically record the various subjects mentioned. Text mining techniques are mainly used for data that is already available in digital format. On the Internet, "text mining" can be used to analyze the content of incoming emails or comments made on forums and social media.

The demand for "data mining" of various types can only increase. If this demand grows, it should have a guiding effect on research in the field of digital, textual and image "data mining" and on the development of viable systems. This is an essential consideration that companies will and must develop. Methods and procedures for retrieving and updating models, with a view to automating decisions and decision-making processes, must be designed at the same time as data storage systems in order to guarantee the interest and usefulness of these systems for the company.

6.6. Conclusion

The "Big Data" phenomenon has changed the way data is managed, as it introduces new issues regarding the volume, transfer speed and type of data. The evolution of the latest technologies such as smartphones, tablets, etc. allow quick access to tools; the Web is becoming a place for data processing thanks to broadband. As mentioned earlier, this phenomenon is characterized by the size or volume of the data. The speed and type of data are also to be considered. In terms of type, Big Data is often attached to unstructured content (web content, customer comments, etc.), which can be a challenge for traditional relational storage and computing environments. Speed refers to the speed with which information is created.

Thanks to various new technologies, it is now possible to analyze and use the large amount of data provided by website log files, social network opinion analysis, streaming videos and environmental sensors. Most data collected in distributed file systems are unstructured types, such as text, images or video. While some of this data may be relatively simple (e.g. calculating simple counts, etc.), others require more complex algorithms that must be developed specifically to work efficiently on the distributed file system.

The quality of this data influences the soundness and relevance of strategic decisions at the enterprise level, hence the need to think "information value chain". All the players in the information value chain are then highly aware of the identification and, above all, the declaration of quality defects; the means to achieve this is the implementation of an internal communication campaign and tools for capturing quality defects in the data. Data governance is a pragmatic approach that formalizes and distributes responsibility for data management across the value chain.

Conclusion

In the world of economic globalization, economic exchanges as well as the rapid and profound transformation of science and technology, modern companies, in order to adapt and remain competitive, must anticipate changes in their environment. As large datasets are currently available from an abundance of sources, companies are looking to use these resources to promote innovation, build customer loyalty and increase operational efficiency. At the same time, they are being challenged for their end use, which requires greater capacity to exploit, analyze and manage the increasing amounts of data forming a "data boom", also known as "Big Data".

The revolutionary dimension of "Big Data" lies in the empowerment of production processes and massive, continuous and ever-faster exchanges of data. Companies collect large amounts of data that arrive via interconnected tools, social networks, emails, surveys, web forms and other data collection mechanisms. Every time someone does a search on Google, uses a mobile application, "tweets" or posts a comment on Facebook, they create a data feed. By simple browsing on the Internet, a user leaves traces of various kinds, which can be retrieved in real-time. It is up to the company to determine which ones may represent a competitive advantage.

These digital traces must be explored through cross-referencing and correlations, obviously using specialized tools, in order to take advantage of them and offer quality products and services, provided that the appropriate development strategies are put in place. So, the "value" that this data represents for the company is added to the different "Vs" of Big Data (volume, variety, velocity, etc.). The subject of the “business value” of data raises many contextual and complex questions and depends very much on the reading that is done (processing, analysis, visualization, etc.) and the importance that is given to it. This is one of the biggest challenges of Big Data, “data mining”.

Businesses need to create value through data flows and content available in real-time. They need to ask themselves whether they are integrating digital in their strategy. This involves transforming data of different forms and sources into “information”, which is complex and often unstructured. But Big Data analysis refers to tools and methods that aim to transform massive amounts of data into strategic information. The challenge is therefore to be able to cross-reference the maximum amount of data from internal and external sources in order to extract the best elements for decision-making.

To create value through the large amounts of data available, it is necessary to:

– develop and create an internal data-friendly culture;

– master structured and unstructured data;

– ensure data quality to guarantee the value of the information, the adaptability, and scalability of the system;

– reinforce initiatives on “Open data” or the provision of data, both public and private, to enable further exploitation;

– work in collaboration between the different departments of the company;

– use visual presentations of data with business intelligence tools, to extract clear and meaningful information for a better understanding of the information;

– to master the techniques of “data scientist” or “massive data analyst” of data a process of knowledge extraction “data mining” or “textmining”;

– make the best decision.

Box C.1. *Creating value from data – suggestions*

Big Data, the processing and analysis of a very large amount of data, must therefore revolutionize the work of companies. However, this approach is not very well developed in French companies. French companies are particularly lagging behind in this area. For two-thirds of them, Big Data is an "interesting concept, but too vague to be a lever for growth". In addition, certain sectors are particularly active in large-scale data collection and processing: this is notably the case for mass retailing, telecoms, media and technology, which are the most mature in terms of the use of customer data in France.

In this sense, France must work quickly to be at the forefront in the perspective of data development. The development of Big Data is the seventh priority of the report "One Principle and Seven Ambitions for Innovation" of the commission "Innovation 2030"[1] chaired by A Lavergeon. The report indicates that the Big Data is one of the main points to be developed and improved in the coming years. According to the same report, the exploitation of the massive data available to companies and public authorities brings new applications and considerable gains in competitiveness.

The report sets out a series of guidelines, such as:

– making public data available for exploitation. France has already launched the "Etalab" project on this point. But the report wants to accelerate the movement according to the British model, to allow start-ups "to create ecosystems in France by developing certain uses for commercial purposes";

– creating five licenses to use the data, depending on whether they come from Pôle emploi, Social Security, National Education and Higher Education, as well as aid for the development of the tourist heritage. The aim here is to strengthen collaboration between public authorities and private actors;

– consecrating a right to experimentation, under the aegis of a "data observatory". The aim is to evaluate the effectiveness of certain practices that allow the exploitation of data, before considering a possible legislative framework;

1 https://www.entreprises.gouv.fr/innovation-2030.

– creating a dedicated technology resource center to remove barriers to entry and reduce the time-to-market for start-ups;

– strengthening the export capacities of SMEs, so as not to remain in the overly landlocked French market, while encouraging the public authorities to intervene more in this sector.

Data has become a gold mine; it represents a raw material like oil in the 20th Century. For geological reasons, oil is concentrated in certain regions of the world, while the data is produced by users all over the world, transported on the web, accumulated and analyzed through specialized techniques and tools. The volume of data produced and to be processed by companies is therefore constantly increasing. This is not a recent finding. But in the age of the Internet and the multiplication of connected devices, the analysis of these warehouses has become complex.

Due to the ever-increasing volume of electronically accessible data, the design and implementation of effective tools to access only relevant information is becoming an absolute necessity. Behind this is the use of expert systems, but also the use of data analysis methods that aim to discover structures and relationships between facts, using elementary data and appropriate mathematical techniques. It is a matter of analyzing these quantities of data in order to find the relevant information to make the right decision.

For example, when you search for or order a book on "Amazon", the site's algorithm suggests other items that represent purchases made by thousands of users before us. "Netflix" goes further: the data collected is used to find out which films to buy rights to from film studios and to develop new programs. "Google maps" offers us a route optimized according to real-time traffic conditions; "Facebook" offers us new friends according to the links already established. All these proposals are formulated through the analysis of a large amount of data collected.

In the United States, the use of an anonymous facial recognition camera at the entrance to nightclubs has been transformed into a service. This is "ScenTap", a mobile application that provides real-time information on the occupancy rate of the premises and the number of men and women present. Managers manage their establishments in real-time and customers can decide where they prefer to go.

In France, the SNCF has focused on developing its business on the Internet and Smartphones, to control the information produced and distributed on the net and to prevent Google from becoming the one to manage customer relations. That is what happened with Booking.com, on which all the hotels are now dependent.

Data is currently produced daily and in real-time, from telephones, credit cards, computers, sensors and so on. It is not only a question of the quantity and speed of production of this data, but also the real revolution lies in what can be created by crossing and analyzing these flows. Big Data processing requires an investment in IT architectures that can store, manage, analyze and visualize huge amounts of data.

This specific technology provides a means to enrich the analysis and show the strength of the available data. You need the right processing method to be able to analyze the data properly. But it is also a question of investing in the human skills to analyze these quantities more precisely and to develop relevant algorithms and models. In other words, the company needs data scientists to make these warehouses talk. It is indeed important to select the most relevant data, and therefore to include in the analysis the notion of quality and not only quantity: this is where the "data governance" strategy comes in.

We can then see that Big Data involves a number of components working together to provide a rich ecosystem and more powerful analyses. Such an ecosystem can be presented as follows.

Big Data opens up new perspectives of knowledge and provides another form of value creation for companies. Today companies have a large amount of data at their disposal. These reservoirs of knowledge need to be explored in order to understand their meaning and to identify the relationships between data and the models explaining their behavior. The interest of Big Data is to be able to take advantage of the data produced by all the actors (companies, individuals, etc.), which become a strategic asset, a tool for value creation and give rise to a new organizational paradigm with possibilities for innovation.

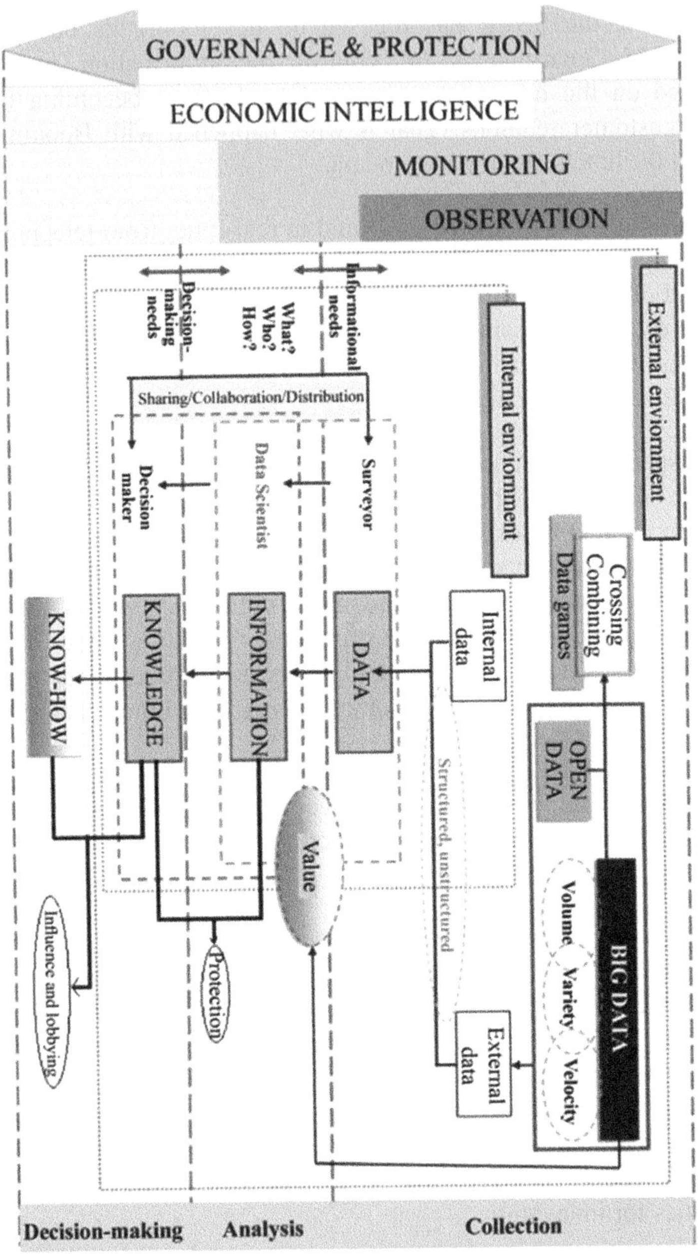

Figure C.1. *The place of data in the digital society: a proposal for modeling through the concept of Economic Intelligence. For a color version of this figure, see www.iste.co.uk/monino/control.zip*

Glossary

To conclude this book, I think it is necessary to recall different definitions of some key words used. Obviously, this will not be exhaustive.

Artificial intelligence: artificial intelligence is a field of science whose goal is to simulate or surpass human intelligence. The problem is that today the term artificial intelligence is attributed to the most basic algorithms that have nothing to do with it, so we have to be careful about abuse.

Brand activism: branding is associated with all the marketing actions carried out by a company in order to create a commercial brand. The objective of branding is to anchor a brand in the consumer's mind. We use the term personal branding for a person's image, personal marketing.

Buzzword: a "buzzword" is a jargon term that is used, for a certain period of time, as a slogan to designate a novelty (technology, product, concept, etc.) and thus draw attention to this novelty (https://www.leslivresblancs.fr/dossier/les-meilleurs-buzzwords-de-2019).

Deepfake: deepfake or hyperfake is an application of deep learning. It is a branch of synthetic media in which a person in an existing image or video is replaced by the likeness of someone else using artificial neural networks. They often combine and superimpose existing media on source media using machine learning techniques known as auto-coding and generating adversary networks.

Deep learning: deep learning is a type of learning derived from the learning machine, which consists of teaching artificial intelligence through the study of examples. These two technologies are subgenres of artificial intelligence.

Influence marketing: influence marketing is surely one of those buzzwords that will eventually become established, already fashionable last year, it's still in the spotlight today. The principle of influence marketing is to pay influencers to promote a product or service.

Mobile first: it consists of designing a site by prioritizing its mobile version, and gradually adapting it for larger screens. This is a display designed primarily for a smartphone version. A mobile first site is created for the mobile and is then adapted to the largest screens. The number of mobile users of web sites and applications is constantly increasing.

The mobile experience has become so important that today Google favors sites with a version adapted to phones.

Nano-influencers: 2018 was the year of micro-influencers, 2019 was the year of nano-influencers, those people who have between 1,000 and 5,000 subscribers on social networks and who are therefore just ordinary people.

Voice search: voice search is very simply defined as a search performed using the voice. It is now present in many homes, particularly in the form of intelligent PDAs such as the Amazon Echo and Google Home.

References

Abiteboul, J., Ghendrih, Ph., Grandgirard, V., Cartier-Michaud, T., Dif-Pradalier, G., Garbet, X., Latu, G., Passeron, C., Sarazin, Y., Strugarek, A., Thomine, O., and ZarzosoHide, D. (2013). Turbulent momentum transport in core tokamak plasmas and penetration of scrape-off layer flows. *Plasma Physics and Controlled Fusion*, 55(7), 074001.

Ackoff, R.L. (1989). From data to wisdom. *Journal of Applied Systems Analysis*, 15, 3–9.

Ageron, B., Goury, M.-L., and Spalanzani, A. (2010). Le Knowledge Management appliqué aux problématiques de développement durable dans la Supply Chain. Centre d'Études et de Recherches Appliquées à la Gestion, 2010–03, E5, Version 1.

Agrell, W. and Huldt, B. (1983). *Clio Goes Spying: Eight Essays on the History of Intelligence*. Scandinavian University Books, Malmö.

Aguilar, F.J. (1967). *Scanning Business Environment*. Macmillan Co., New York.

Albaum, G. (1962). A new approach to the information function in marketing, Theses, University of Wisconsin. *Communication*, 16(2), 163–176.

Albert, P., Bernasconi, M., and Gaynor, L. (April, 2002). Les incubateurs: émergence d'une nouvelle industrie, comparaison des acteurs et de leurs stratégies: France, Allemagne, Royaume-Uni, Etats-Unis, CERAM, DIGITP.

Almeida, F. (2012). Web 2.0 technologies and social networking security fears in enterprises. *International Journal of Advanced Computer Science and Applications*, 3(2), 152–156.

Amabile, T.M. (2002). Motiver la créativité dans les organisations: faites ce que vous aimez et aimez ce que vous faites. In *Créativité organisationnelle: regards sur l'individu, l'entreprise et l'economie*, Getz, I. (ed.). Éditions Vuibert, Paris.

Bastien, L. (2019). Base de données: qu'est-ce que c'est? Définition et présentation [Online]. Available: https://www.lebigdata.fr/base-de-donnees#ftoc-heading-1.

Baumard, P. (1991). *Stratégie et surveillance des environnements concurrentiels*. Masson, Paris.

Baumard, P. (1992). *Guerre économique et communauté d'intelligence*. La Revue Politique et Parlementaire, Paris.

Baumard, P. (March, 1999). La guerre du savoir a commencé. *L'Expansion Management Review*, 92, 60–69.

Baumard, P. and Benvenuti, J.A. (1998). *Compétitivité et systèmes d'information, de l'outil d'analyse au management stratégique*. InterEditions, Paris.

Benkaraache, T. and Benabdelouahed, R. (2015). L'émergence du Big Data dans le secteur bancaire: cas des banques marocaines. *VSST 2015 Congress Proceedings*, Granada, Spain.

Benkaraache, T., Monino, J.-L., and Salam, G. (2015). Intégration du Big Data dans le Supply Chain: vers une supplychain 'intelligente'. *AIM 2015 Proceedings*, Rabat, Morocco [Online]. Available: http://aim.asso.fr/index.php/mediatheque/viewcategory/34-aim-2015?start=60.

Berners-Lee, T. (2000). Entretien, Courier de l'UNESCO.

Berners-Lee, T., Hendler, J., and Lassila, O. (2001). The Semantic Web [Online]. Available: http://www.sciam.com/article.cfm?id=the-semantic-web.

Berry, M.J.A. and Linoff, G.S. (2000). *Mastering Data Mining*. Wiley, New York.

Boya-Guierdel, C. (2008). Modélisation de la relation entre les informations exogènes médiatiques et le marché des actions. PhD thesis, Montpellier University, France.

Boya-Guierdel, C. (2009). Analyse des informations exogènes secondaires et mesure d'impact sur le marché action. *Revue internationale d'intelligence économique*, 1, 237–253.

Brown, B., Manyika, J., and Chui, M. (2011). Are you ready for the era of 'big data'? [Online]. Available: https://www.mckinsey.com/business-functions/strategy-and-corporate-finance/our-insights/are-you-ready-for-the-era-of-big-data.

Bughin, J. and Chui, M. (2010). The rise of the networked enterprise: Web 2.0 finds its payday [Online]. Available: https://www.mckinsey.com/industries/technology-media-and-telecommunications/our-insights/the-rise-of-the-networked-enterprise-web-20-finds-its-payday [Accessed 2 February 2014].

Bughin, J., Manyika, J., and Chui, M. (2010). Clouds, big data, and smart assets: Ten tech-enabled business trends to watch. *McKinsey Quarterly*, 56(1), 75–86.

Bulinge, F. (2002). Pour une culture de l'information dans les PMO: un modèle incrémental d'intelligence économique. PhD Thesis, University of Toulon, France.

Bulinge, F. (2006). Le cycle du renseignement: analyse critique d'un modèle empirique. *Revue internationale des sciences sociales Marketing & Communication*, 3(3), 36–52.

Bulinge, F. (2007). Un modèle d'analyse collective en situation: la war-room. *Marketing & Communication*, 7(4), 50–69.

Bulinge, F. and Agotinelli, S. (2005). L'analyse d'information: d'un modèle individuel à une culture collective. Management et communication pour une économie de la connaissance.

Butler, M. (2009). The business value of business intelligence: A review of benefits, costs and risks. *Business Value Series*, 2, 4 and 7.

Cavagna, E., Dumazert, J.P., and Monino, J.-L. (2004). Construction d'un indice de satisafction: une réfexion autour de la question de la mesure. *Revue Française de Marketing*, 197, 61–77.

Chesnais, F. (1995). La montée des firmes-réseaux. *Alternatives Economiques*, Special edition 23.

Chignard, S. (2012). *Open data: comprendre l'ouverture des données publiques*. FYP, Limoges, France.

Chui, M. (2013). The social economy: Unlocking value and productivity through social technologies. *McKinsey*, 153.

Chui, M., Miller, A., and Roberts, R.P. (2009). Six ways to make Web 2.0 work: Web 2.0 tools present a vast array of opportunities – for companies that know how to use them. *The McKinsey Quarterly*, 0–7 [Online]. Available: http://www.mckinsey.com/insights/business_technology/six_ways_to_make_web_20_work.

Cleveland, W. (2014). Data science: An action plan for expanding the technical areas of the field of statistics [Online]. Available: http://www.stat.purdue.edu/~wsc/papers/datascience.pdf.

Cody, W.F., Kreulen, J.T., Krishna, V., and Spangler, W.S. (2002). The integration of business intelligence and knowledge management. *IBM Systems Journal*, 41(4), 1 and 3.

Communication de la Commission au Parlement européen, au Conseil, au Comité économique et social européen et au Comité des Régions (2011). L'ouverture des données publiques: un moteur pour l'innovation, la croissance et une gouvernance transparente, 12 December 2012.

Cortes, C. and Pregibon, D. (July 6–11, 1997). Mega-monitoring, unpublished paper presented at the University of Washington/Microsoft Summer Research Institute on Data Mining.

Coudercy, L., Delobelle, G., from ONEMA and BRGM respectively (2014). Les données cartographiques dans une démarche Open Data: Cas des données sur l'eau, première rencontre CEREMA-ONEMA: eaux et milieux aquatiques, vers de nouvelles synergies, Bron, December 2014.

Crozet, Y. and Musso, P. (2003). *Networks, services and territories, horizon 2020*. Editions De L'aube.

David, A. (2006). *L'impact des TIC: logistique, transport, relation de service, organisation*. La Documentation française, Paris.

Dupuis, A. (2004). Gouvernance, ajustement mutuel, communautés humaines et intelligence collective, CERGO research notebook.

Dupuy G. and Geneau De Lamarliere I. (2007). *Nouvelles échelles des firmes et des réseaux, un défi pour l'aménagement*. L'Harmattan, Paris.

Edelstein, H.A. (1999). *Introduction to Data Mining and Knowledge Discovery*, 3rd edition. Two Crows Corp, Potomac, MD.

Eliot, T.S. (1934). The Rock, Faber & Faber. Available: http://www.wisdomportal.com/Technology/TSEliot-TheRock.html.

Ermine, J.-L., Moradi, M., and Brunel, S. (2012). Une chaîne de valeur de la connaissance. *Gestion des connaissances dans la société et les organisations*, special issue, 16, 29–40.

Escarpit, R. (1976). *Théorie générale de l'information et de la communication*. Classiques Hachette, Paris.

European Commission (2012). EC ISA Case Study: How Linked Data is transforming e-Government.

Falquet, G. and Mottaz Jiang, C.-L. (2001). Navigation hypertexte dans une ontologie multi-point de vue, *NîmesTIC 2001 Proceedings*, Nîmes, France.

Farbey, B., Targett, D., and Land, F. (1995). Evaluating business information systems: Reflections on an empirical study. *Revue Information Systems Journal*, 5(4), 242.

Fayyad, U.M., Piatetsky-Shapiro, G., Smyth, P., and Uthurusamy, R. (1996). *Advances in Knowledge Discovery & Data Mining*. MIT Press, Cambridge, MA.

Fioretti, M. and Sant'Anna, S. (2010). Pisa "Open Data Open Society", report in the framework of the European network of DIME excellence (Dynamics of Institutions and Markets in Europe) [Online]. Available: http://www.lem.sssup.it/WPLem/odos/odos.html.

Flici, A. (2008). A systematic evaluation of business intelligence tools: An investigation into potential tangible/intangible value, PhD Thesis, Brunel University research paper, pp. 2–3.

Foray, D. (2000). *L'économie de la connaissance*. La Découverte, Paris.

Franco, J.-M. (1997). *Le Data Warehouse et le Data Mining*. Eyrolles, Paris.

Gardarin, G. (1999). *Internet/intranet et bases de données*. Eyrolles, Paris.

Getz, I., Robinson, A.G., and Leroy, D. (2007). *Vos idées changent tout!* Editions d'Organisation, Paris.

Goëta, S. (2012). Open data: qu'ouvre-t-on avec les données publiques? CELSA Thesis, prepared under the direction of Richard V.

Groupe de travail sur les données ouvertes (GTDO) (2011). Rapport sur l'ouverture des données de la Ville de Montréal: Un capital numérique générateur d'innovation et de participation [Online]. Available: http://www1.ville.montreal.qc.ca/banque311/webfm_send/1453.

Guyot, B. (1999). Le système d'information comme objet de recherche. In *Communication organisationnelles, objets, pratiques, dispositifs*, Delcambre, P. (ed.). Presses universitaires de Rennes, Rennes.

Han, J. and Kamber, M. (2000). *Data Mining: Concepts and Techniques*. Morgan-Kaufman, New York.

Harbulot, C. (1990). *Techniques offensives et guerre économique*, Étude No 131, Centre de Prospective et d'Évaluation/Aditech, Paris.

Hastie, T., Tibshirani, R., and Friedman, J.H. (2001). *The Elements of Statistical Learning: Data Mining, Inference, and Prediction*. Springer, New York.

Heath, T. and Bizer, C. (2007). *Web de données "Méthodes et outils pour les données liées"*. Pearson, France.

Hopkins, B. and Evelson, B. (2011). Expand your digital horizon with Big Data, Forrester research [Online]. Available: http://www.asterdata.com/newsletter-images/30-04-2012/resources/Forrester_Expand_Your_Digital_Horiz.pdf?lf1=97073534d99767367906a1885620.

Jacob, R. and Turcot, S. (2000). La PME apprennante, information, connaissance, interaction, intelligence. Monitoring report presented at the Développement Economique Canada.

Jakobiak, F. (2009). *L'intelligence économique: Techniques et outils*, 2nd edition. Editions d'Organisation, Paris.

Jarrosson, B. (1994). *Décider ou ne pas décider: réflexions sur les processus de la décision*. Maxima, Paris.

Juillet, A. (2004). Du renseignement à l'intelligence économique. Défense Nationale, C. Quesnot.

Julien, P.A., Vaghely, I., and Carrier, C. (2003). PME et contrôle de l'information: le rôle du troisième homme. *Colloque d'Agadir*.

Kefi, H. and Kalika, M. (2004). *Evaluation des systèmes d'information: une perspective organisationnelle*. Editions Economica.

Lacombe, R., Bertin, P.-H., Vauglin, F., and Viellefosse, A. (2011). Pour une politique ambitieuse des données publiques. Report, French Minister for Industry, Energy and Digital Economy [Online]. Available: http://www.ladocumentationfrancaise.fr/var/storage/rapports-publics/114000407/0000.pdf.

Lefébure, R. and Venturi, G. (1998). *Le data Mining*. Eyrolles, Paris.

Le Guen, F. (2013). *Analyse et visualisation des données*. Pearson, France.

Lesca, H. (1986). *Système d'information pour le management stratégique de l'entreprise*. McGraw-Hill, Paris.

Lesca, H. (2008). Gouvernance d'une organisation: prévoir ou anticiper? *La Revue des sciences de gestion Direction et Gestion*, 231–232, 11–17.

Les Décodeurs (2017). Décodex: qu'est-ce qu'une source d'information? [Online]. Available: https://www.lemonde.fr/les-decodeurs/article/2017/01/23/decodex-qu-est-ce-qu-une-source-d-information_5067722_4355770.html.

Luhn, H.P. (1957). The Automatic Creation of Literature Abstracts. *IBM Journal*, April, 159–165.

Malthus, T. (1798). An Essay on the Principle of Population [Online]. Available: http://www.esp.org/books/malthus/population/malthus.pdf.

Manyika, J, McKinsey Global Institute, Chui, M., Brown, B., Bughin, J., Dobbs, R., Roxburgh, C., and Byers, A.H. (2011). Big data: The next frontier for innovation, competition, and productivity. Report, McKinsey Global Institute.

Martre, H. (ed.) (1994). *Intelligence économique et stratégie des entreprises*. Commissariat Général du Plan, La Documentation française, Paris.

Matouk, J. (2014). La Revue de la MTRL. *Mutuelle et Santé*, no. 84.

Matouk, J. and Monino, J.-L. (2005). Le Marché de Paris à la mémoire courte. *Revue d'économie financière*, 81, 133–155.

Mayère, A. (1990). Comment définir l'information? In *Pour une économie de l'information*, Mayère, A. (ed.). CNRS Editions, Paris.

Mayer-Schonberger, V. and Cukier, K. (2013). *Big Data: A Revolution That Will Transform How We Live, Work and Think*. John Murray, London.

McAfee, A.P. (2006). Enterprise 2.0: The dawn of emergent collaboration. *MIT Sloan Management Review*, 47(3), 20–29.

McMahon, A., Liu, Y., Crossland, R., Brown, D., Leal, D., and et Devlukia, J. (2004). A best practice advice system to support automotive engineering analysis processes. *Engineering with Computers*, 19(4), 271–283.

McNulty, E. (2014). Big Data: The Seven V's [Online]. Available: http://dataconomy.com/seven-vs-big-data/.

Mercier, D. (2014). Données ouvertes: qualité et visualisation [Online]. Available: http://dianemercier.quebec/donnees-ouvertes-qualite-et-visualisation/.

Moinet, N. (2009). Mythes et réalités de l'intelligence économique. Analyse d'une dynamique à la recherche de ses concepts opératoires & perspectives de recherche en communication, HDR Thesis.

Monino, J.-L. (2012a). Un modèle hiérarchique pour l'Intelligence économique, *Colloque Intelligence économique compétitivité et attractivité des territoires*, Ibn TOFAIL University and Montpellier University – Sud de France.

Monino, J.-L. (2012b). L'information au coeur de l'intelligence économique stratégique. *Réseau de Recherche sur l'Innovation*, 27, working papers.

Monino, J.-L. (2013). L'information au cœur de l'intelligence économique stratégique. In *Intelligence, entreprises et territoires*, Monino, J.-L. (ed.). L'Harmattan, Paris.

Monino, J.-L. (2015). Big Data, Open Data et valorization des données, *Research network on innovation*, working papers, 48.

Monino, J.-L. (2016). Data value, Big Data analytics, and decision-making. *Journal of the Knowledge Economy*, Carayannis, E.G. (ed.). Springer Science and Business Media LLC. DOI: 10.1007/s13132-016-0396-2.

Monino, J.-L. and Boussetta, M. (2013). Introduction générale, *Intelligence, entreprises et territoires*. L'Harmattan, Paris.

Monino, J.-L. and Boya, C. (2010). The impact of exogenous information on stock value through the coloration concept: A test model. *Journal of Innovation Economics*, 6(2), 163–180.

Monino, J.-L. and Boya, C. (2011). La coloration de l'information dans l'efficience semi forte. *Revue Innovations*, 36(3), 147–157.

Monino, J.-L. and Boya, C. (2013). Modélisation non paramétrique de la relation entre les séries: la cointégration qualitative. *Innovations*, De Boeck Supérieur, Paris.

Monino, J.-L. and Boya, C. (2016). Un modèle d'usure du pouvoir: l'approche endogène et exogène par l'information. *Cahier du CEDIMES*.

Monino, J.-L. and Cavagna, E. (1999). Indice de satisfaction: conceptualisation et mise en application dans le cadre de la Chambre de Commerce et d'Industrie de Montpellier. *Revue Modulad*, INRIA, no. 22.

Monino, J.-L. and Fechtali, T. (2017). De la donnée à l'information: une transformation essentielle pour créer de la valeur. *Presentation WIM 2017 colloquium*, Ibn Tofai University, Kenitra, Morocco.

Monino, J.-L. and Lucato, G. (2005). Manager l'information: comment faire aujourd'hui? Les matins de la cité, La CCI de Montpellier.

Monino, J.-L. and Sedkaoui, S. (2013a). L'algorithme de l'escargot: Un exemple de saisir la fenêtre d'opportunité pour stimuler Big Data. In *Distributed Computing and Artificial Intelligence*, Omatu, S., Neves, J., Rodriguez, J.M.C., Paz Santana, J.F., and Rodriguez González, S. (eds). Springer, Cham.

Monino, J.-L. and Sedkaoui, S. (2013b). Les TIC un outil indispensable pour une démarche d'intelligence économique. *Marché et organisations*, 18(2), 173–188.

Monino, J.-L. and Sedkaoui, S. (2016). *Big Data, Open Data and Data Development*. ISTE Ltd, London and Wiley, Hoboken, NJ.

Monino, J.-L., Sedkaoui, S., and Matouk, J. (2014). Big data, éthique des données, et entreprises. *Les Cahiers du CEDIMES*, 8(2) [Online]. Available: http://cedimes.com/images/documents/cahiers_cedimes/cahier_cedimes_vol_8_N2_2014.pdf.

OECD (2013). Exploring data-driven innovation as a new source of growth: Mapping the policy issues raised by "Big Data", OECD Digital Economy Papers, no. 222, OECD Publishing.

Peugeot, V. (2010). Les enjeux publics économiques et citoyens de l'ouverture des données: l'expérience britannique, Orange Labs [Online]. Available: http://laborange.academia.edu/Val%C3%A9riePeugeot/Papers/343143/Les_enjeux_publics_economiques_et_citoyens_de_louverture_des_donnees_lexperience_britannique.

Porter, M. (March-April 1979). How competitive forces shape strategy. *Harvard Business Review* [Online]. Available: http://www.atelier-intelligence-economique.fr/index.php/quest-ce-que-lintelligence-economique/.

Pimor, Y. and Fender, M. (2008). *Logistique: production, distribution, soutien*. Dunod, Paris.

Pool, I.S. (1984). *Communications Flows: A Census in the United States and Japan*. Elsevier Science, New York.

Rotella, P. (2012). Is Data The New Oil? [Online]. Available: http://www.forbes.com/sites/perryrotella/2012/04/02/is-data-the-new-oil/.

Saïd, A. (2004). Rôle de l'intelligence économique dans la stratégie de l'entreprise. PhD Thesis, Mention Sciences de Gestion.

Salmon, R. and de Linarès, Y. (1997). *L'intelligence compétitive, une combinaison subtile pour gagner ensemble*. Economica, Paris.

Taylor, R.S. (October 5–10, 1980). Value-added aspects of the information process. *Communicating Information: Proceedings of the 43rd, ASIS Annual Meeting*, Vol. 17, Anaheim, California.

Thepaut, Y. (2002). *Pouvoir, Information, Economie*. Economica, Paris.

Tufféry, S. (2005). *Data Mining et statistique décisionnelle*. Technip, Paris.

Vickery, G. (23 September, 2011). *Review of Recent Studies on PSI Re-use and Related Market Developments*. Information Economics, Paris, 2011.

Wilensky, H. (1967). *Organizational Intelligence: Knowledge and Policy in Government and Industry*. Basic Books, New York.

Zara, O. (2008). *Le management de l'intelligence collective, vers une nouvelle gouvernance*, 2nd edition. M21 éditions.

Webography

http://bbf.enssib.fr/

https://www.contrefacon-riposte.info/technologies/4423-l-universite-de-montpellier-pionniere-dans-la-lutte-contre-la-contrefacon-des-diplomes http://data.gouv.fr

http://www.data-publica.com/

http://www.datazen.com/

http://donneesouvertes.info/

https://www.etalab.gouv.fr/blog

http://www.gapminder.org

http://www.guardian.co.uk/open-platform

http://www.handimap.org/

http://www.indeed.com/jobtrends/Data-scientist.html

http://www.lecfomasque.com/

http://www.leparisien.fr/startup/

http://www.mckinsey.com/insights/mgi

http://www.myfrenchstartup.com/

http://owni.fr

https://powerbi.microsoft.com/

http://www.redsen-consulting.com/

https://support.office.com/

BFM Business (2009). Le web intelligent prend forme [Online]. Available: https://bfmbusiness.bfmtv.com/01-business-forum/le-web-intelligent-prend-forme-506931.html.

Business Herald (2020). Yannick Poivey: Big Data, quelles conséquences pour les prestataires en intelligence économique? [Online]. Available: http://business-herald.com/enquete/yannick-poivey-big-data-consequences-en-intelligence-economique.

CaddE-Réputation (2014). Petite bibliographie critique sur les Big Data [Online]. Available: http://cadderep.hypotheses.org/94/.

CCM (2011). Web sémantique: quelles applications aujourd'hui? [Online]. Available: http://www.commentcamarche.net/faq/29732-web-semantique-quelles-applications-aujourd-hui.

Chignard, S. (2011). Comment évaluer l'impact économique de l'open data au niveau local? [Online]. Available: http://fr.slideshare.net/lacantinerennes/open-data-garage-comment-valuer-limpact-conomique-de-lopen-data-local.

Cointot, J.-C. and Eychenne, Y. (2014). La Révolution Data [Online]. Available: http://www.dunod.com/entreprise-economie/entreprise-et-management/strategie-et-politique-de-lentreprise/ouvrages-professionnels/la-revoluti-1.

Dèbes, F. (2013). Créneau porteur: 5 exemples de start-up open data [Online]. Available: http://business.lesechos.fr/entrepreneurs/idees-de-business/creneau-porteur-5-exemples-de-start-up-open-data-57178.php?dxC0WjY1PjomkZJg.99.

Decideo (2012). Le Big Data: un cadeau empoisonné pour les Data Miners? [Online]. Available: http://www.decideo.fr/Le-Big-Data-un-cadeau-empoisonne-pour-les-Data-Miners_a5457.html.

Decider Ensemble (2018). Actualités [Online]. Available: https://www.deciderensemble.com/articles/25343-methanisation-mise-en-place-un-dialogue-territorial.

Diebold, F.X. (2019). On the origin(s) and development of "Big Data": The phenomenon, the term, and the discipline [Online]. Available: http://www.ssc.upenn.edu/~fdiebold/papers/paper112/Diebold_Big_Data.pdf.

Données Ouvertes (2012). Où est le marché de l'open data? [Online]. Available: http://donneesouvertes.info/2012/03/12/ou-est-le-marche-de-lopen-data/.

Grodziski, J. (2011). Quel sens donner aux mots "Données", "Informations", "Connaissances" et "Compétences"? [Online]. Available: http://www.redsen-consulting.com/2011/11/quel-sens-donner-aux-mots-donnees-informations-connaissances-et-competences/.

Haehnsen, E. (2013). L'Open Data est à la mode et… cherche son modèle [Online]. Available: http://www.latribune.fr/technos-medias/informatique/20130208trib000747692/l-open-data-est-a-la-mode-et-cherche-son-modele.html.

ITExpert. DSI et Big Data – développez la proximité avec le marketing et embarquez le dans la stratégie data [Online]. Available: http://www.it-expertise.com/dsi-et-big-data-developpez-la-proximite-avec-le-marketing-et-embarquez-le-dans-la-strategie-data/.

Jobillico (2014). Un métier d'avenir: le scientifique des données [Online]. Available: http://www.jobillico.com/blog/metier-avenir-scientifique-des-donnees/.

JOUVE (2019). L'engagement de la France pour une santé plus digitale doit se traduire désormais dans les actes [Online]. Available: http://www.jouve.com/fr/la-data-visualisation-dataviz-dans-tous-ses-etats.

Lemonade Informatique (2020). Start-up [Online]. Available: http://www.lemondeinformatique.fr/start-up-128.html.

Les Clés de Demain (2019). A qui profite l'open data? [Online]. Available: http://lesclesdedemain.lemonde.fr/business/a-qui-profite-l-open-data-_a-56-1411.html.

McNulty, E. (2014). Understanding Big Data: The seven V's [Online]. Available: http://dataconomy.com/seven-vs-big-data/.

Microsoft (2020a). Featured stories [Online]. Available: http://www.microsoft.com/casestudies/Case_Study_Detail.aspx?casestudyid=710000003034.

Microsoft (2020b). Power Pivot: analyse et modélisation de données puissantes dans Excel [Online]. Available: https://support.office.com/fr-fr/article/Compl%C3%A9ment-PowerPivot-a9c2c6e2-cc49-4976-a7d7-40896795d045?CorrelationId=4299016a-384e-4367-842e-a5faab3d3384&ui=fr-FR&rs=fr-FR&ad=FR.

Open Bibliography and Open Bibliographic Data (2014). Principes des données bibliographiques ouvertes [Online]. Available: http://openbiblio.net/principles/fr/.

OpenDataFrance (2020). Observatoire open data des territoires [Online]. Available: http://opendatafrance.net/category/data-publica/.

Orange. [Online]. Available: http://www.orange.com/fr_FR/presse/communiques/cp111201fr.jsp.

Pasquesoone, V. (2012). En France, l'open data en marche [Online]. Available: http://www.lemonde.fr/politique/article/2012/05/31/en-france-l-open-data-en-marche_1709874_823448.html.

Porter, M. (1979). How Competitive Forces Shape Strategy, Harvard Business Review, March-April [Online]. Available: http://www.atelier-intelligence-economique.fr/index.php/quest-ce-que-lintelligence-economique/.

Regards Citoyens (2017). Open Data en France [Online]. Available: https://www.regardscitoyens.org/open-data-en-france/.

Robert Half (2020). Blog – conseils emploi [Online]. Available: http://www.roberthalf.fr/id/PR-03954/jobs-en-or-2014.

Scoop It (2019). OpenDataFrance [Online]. Available: https://www.scoop.it/u/open-data-france/curated-scoops.

Silvae (2011). Bibliothèques publiques et open data: quels enjeux? [Online]. Available: http://www.bibliobsession.net/2011/06/07/bibliotheques-publiques-et-donnees-ouvertes/.

Spivack, N. (2004). New version of my "Metaweb" graph – The future of the net [Online]. Available: http://novaspivack.typepad.com/nova_spivacks_weblog/2004/04/new_version_of_.html.

Statistica (2016). Data mining [Online]. Available: http://www.statsoft.fr/concepts-statistiques/glossaire/d/data.html.

Transport Intelligent June 2, 2017 - Sector of Excellence in France - Intelligent Transport Systems and Services (ITS). [Online]. Available: https://www.ecologique-solidaire.gouv.fr/sites/default/files/14085_Transport-intelligent_Web_planches_FR.pdf.

W3C (2017). SweoIG/TaskForces/CommunityProjects/LinkingOpenData [Online]. Available: http://esw.w3.org/topic/SweoIG/TaskForces/CommunityProjects/LinkingOpenData.

WOD Discovery (2012). Linked Open Data star scheme by example [Online]. Available: http://lab.linkeddata.deri.ie/2010/star-scheme-by-example.

Index

K, M, N, P

Q, S, T,V

Other titles from

iSTE

in

Innovation, Entrepreneurship and Management

2020

ANDREOSSO-O'CALLAGHAN Bernadette, DZEVER Sam, JAUSSAUD Jacques, TAYLOR Richard
Sustainable Development and Energy Transition in Europe and Asia
(Innovation and Technology Set – Volume 9)

CORON Clotilde
Quantifying Human Resources: Uses and Analysis
(Technological Changes and Human Resources Set – Volume 2)

CORON Clotilde, GILBERT Patrick
Technological Change
(Technological Changes and Human Resources Set – Volume 1)

CERDIN Jean-Luc, PERETTI Jean-Marie
The Success of Apprenticeships: Views of Stakeholders on Training and Learning
(Human Resources Management Set – Volume 3)

DIDAY Edwin, GUAN Rong, SAPORTA Gilbert, WANG Huiwen,
Advances in Data Science
(Big Data, Artificial Intelligence and Data Analysis Set – Volume 4)

DOS SANTOS PAULINO Victor
Innovation Trends in the Space Industry
(Smart Innovation Set – Volume 25)

GUILHON Bernard
Venture Capital and the Financing of Innovation
(Innovation Between Risk and Reward Set – Volume 6)

MAKRIDES Andreas, KARAGRIGORIOU Alex, SKIADAS Christos H.
Data Analysis and Applications 3: Computational, Classification, Financial, Statistical and Stochastic Methods
(Big Data, Artificial Intelligence and Data Analysis Set – Volume 5)
Data Analysis and Applications 4: Financial Data Analysis and Methods
(Big Data, Artificial Intelligence and Data Analysis Set – Volume 6)

MASSOTTE Pierre, CORSI Patrick
Complex Decision-Making in Economy and Finance

MORLAT Clément
Sustainable Productive System: Eco-development versus Sustainable Development
(Smart Innovation Set – Volume 26)

SAULAIS Pierre, ERMINE Jean-Louis
Knowledge Management in Innovative Companies 2: Understanding and Deploying a KM Plan within a Learning Organization
(Smart Innovation Set – Volume 27)

2019

AMENDOLA Mario, GAFFARD Jean-Luc
Disorder and Public Concern Around Globalization

BARBAROUX Pierre
Disruptive Technology and Defence Innovation Ecosystems
(Innovation in Engineering and Technology Set – Volume 5)

DOU Henri, JUILLET Alain, CLERC Philippe
Strategic Intelligence for the Future 1: A New Strategic and Operational Approach
Strategic Intelligence for the Future 2: A New Information Function Approach

FRIKHA Azza
Measurement in Marketing: Operationalization of Latent Constructs

FRIMOUSSE Soufyane
Innovation and Agility in the Digital Age
(Human Resources Management Set – Volume 2)

GAY Claudine, SZOSTAK Bérangère L.
Innovation and Creativity in SMEs: Challenges, Evolutions and Prospects
(Smart Innovation Set – Volume 21)

GORIA Stéphane, HUMBERT Pierre, ROUSSEL Benoît
Information, Knowledge and Agile Creativity
(Smart Innovation Set – Volume 22)

HELLER David
Investment Decision-making Using Optional Models
(Economic Growth Set – Volume 2)

HELLER David, DE CHADIRAC Sylvain, HALAOUI Lana, JOUVET Camille
The Emergence of Start-ups
(Economic Growth Set – Volume 1)

HÉRAUD Jean-Alain, KERR Fiona, BURGER-HELMCHEN Thierry
Creative Management of Complex Systems
(Smart Innovation Set – Volume 19)

LATOUCHE Pascal
Open Innovation: Corporate Incubator
(Innovation and Technology Set – Volume 7)

LEHMANN Paul-Jacques
The Future of the Euro Currency

LEIGNEL Jean-Louis, MÉNAGER Emmanuel, YABLONSKY Serge
Sustainable Enterprise Performance: A Comprehensive Evaluation Method

LIÈVRE Pascal, AUBRY Monique, GAREL Gilles
Management of Extreme Situations: From Polar Expeditions to Exploration-Oriented Organizations

MILLOT Michel
Embarrassment of Product Choices 2: Towards a Society of Well-being

N'GOALA Gilles, PEZ-PÉRARD Virginie, PRIM-ALLAZ Isabelle
Augmented Customer Strategy: CRM in the Digital Age

NIKOLOVA Blagovesta
The RRI Challenge: Responsibilization in a State of Tension with Market Regulation
(Innovation and Responsibility Set – Volume 3)

PELLEGRIN-BOUCHER Estelle, ROY Pierre
Innovation in the Cultural and Creative Industries
(Innovation and Technology Set – Volume 8)

PRIOLON Joël
Financial Markets for Commodities

QUINIOU Matthieu
Blockchain: The Advent of Disintermediation

RAVIX Joël-Thomas, DESCHAMPS Marc
Innovation and Industrial Policies
(Innovation between Risk and Reward Set – Volume 5)

ROGER Alain, VINOT Didier
Skills Management: New Applications, New Questions
(Human Resources Management Set – Volume 1)

SAULAIS Pierre, ERMINE Jean-Louis
Knowledge Management in Innovative Companies 1: Understanding and Deploying a KM Plan within a Learning Organization
(Smart Innovation Set – Volume 23)

SERVAJEAN-HILST Romaric
Co-innovation Dynamics: The Management of Client-Supplier Interactions for Open Innovation
(Smart Innovation Set – Volume 20)

SKIADAS Christos H., BOZEMAN James R.
Data Analysis and Applications 1: Clustering and Regression, Modeling-estimating, Forecasting and Data Mining
(Big Data, Artificial Intelligence and Data Analysis Set – Volume 2)
Data Analysis and Applications 2: Utilization of Results in Europe and Other Topics
(Big Data, Artificial Intelligence and Data Analysis Set – Volume 3)

VIGEZZI Michel
World Industrialization: Shared Inventions, Competitive Innovations and Social Dynamics
(Smart Innovation Set – Volume 24)

2018

BURKHARDT Kirsten
Private Equity Firms: Their Role in the Formation of Strategic Alliances

CALLENS Stéphane
Creative Globalization
(Smart Innovation Set – Volume 16)

CASADELLA Vanessa
Innovation Systems in Emerging Economies: MINT – Mexico, Indonesia, Nigeria, Turkey
(Smart Innovation Set – Volume 18)

CHOUTEAU Marianne, FOREST Joëlle, NGUYEN Céline
Science, Technology and Innovation Culture
(Innovation in Engineering and Technology Set – Volume 3)

CORLOSQUET-HABART Marine, JANSSEN Jacques
Big Data for Insurance Companies
(Big Data, Artificial Intelligence and Data Analysis Set – Volume 1)

CROS Françoise
Innovation and Society
(Smart Innovation Set – Volume 15)

DEBREF Romain
Environmental Innovation and Ecodesign: Certainties and Controversies
(Smart Innovation Set – Volume 17)

DOMINGUEZ Noémie
SME Internationalization Strategies: Innovation to Conquer New Markets

ERMINE Jean-Louis
Knowledge Management: The Creative Loop
(Innovation and Technology Set – Volume 5)

GILBERT Patrick, BOBADILLA Natalia, GASTALDI Lise,
LE BOULAIRE Martine, LELEBINA Olga
Innovation, Research and Development Management

IBRAHIMI Mohammed
Mergers & Acquisitions: Theory, Strategy, Finance

LEMAÎTRE Denis
Training Engineers for Innovation

LÉVY Aldo, BEN BOUHENI Faten, AMMI Chantal
Financial Management: USGAAP and IFRS Standards
(Innovation and Technology Set – Volume 6)

MILLOT Michel
Embarrassment of Product Choices 1: How to Consume Differently

PANSERA Mario, OWEN Richard
Innovation and Development: The Politics at the Bottom of the Pyramid
(Innovation and Responsibility Set – Volume 2)

RICHEZ Yves
Corporate Talent Detection and Development

SACHETTI Philippe, ZUPPINGER Thibaud
New Technologies and Branding
(Innovation and Technology Set – Volume 4)

SAMIER Henri
Intuition, Creativity, Innovation

TEMPLE Ludovic, COMPAORÉ SAWADOGO Eveline M.F.W.
Innovation Processes in Agro-Ecological Transitions in Developing Countries
(Innovation in Engineering and Technology Set – Volume 2)

UZUNIDIS Dimitri
Collective Innovation Processes: Principles and Practices
(Innovation in Engineering and Technology Set – Volume 4)

VAN HOOREBEKE Delphine
The Management of Living Beings or Emo-management

2017

AÏT-EL-HADJ Smaïl
The Ongoing Technological System
(Smart Innovation Set – Volume 11)

BAUDRY Marc, DUMONT Béatrice
Patents: Prompting or Restricting Innovation?
(Smart Innovation Set – Volume 12)

BÉRARD Céline, TEYSSIER Christine
Risk Management: Lever for SME Development and Stakeholder Value Creation

CHALENÇON Ludivine
Location Strategies and Value Creation of International Mergers and Acquisitions

CHAUVEL Danièle, BORZILLO Stefano
The Innovative Company: An Ill-defined Object
(Innovation between Risk and Reward Set – Volume 1)

CORSI Patrick
Going Past Limits To Growth

D'ANDRIA Aude, GABARRET Inés
Building 21st Century Entrepreneurship
(Innovation and Technology Set – Volume 2)

DAIDJ Nabyla
Cooperation, Coopetition and Innovation
(Innovation and Technology Set – Volume 3)

FERNEZ-WALCH Sandrine
The Multiple Facets of Innovation Project Management
(Innovation between Risk and Reward Set – Volume 4)

FOREST Joëlle
Creative Rationality and Innovation
(Smart Innovation Set – Volume 14)

GUILHON Bernard
Innovation and Production Ecosystems
(Innovation between Risk and Reward Set – Volume 2)

HAMMOUDI Abdelhakim, DAIDJ Nabyla
Game Theory Approach to Managerial Strategies and Value Creation
(Diverse and Global Perspectives on Value Creation Set – Volume 3)

LALLEMENT Rémi
Intellectual Property and Innovation Protection: New Practices and New Policy Issues
(Innovation between Risk and Reward Set – Volume 3)

LAPERCHE Blandine
Enterprise Knowledge Capital
(Smart Innovation Set – Volume 13)

LEBERT Didier, EL YOUNSI Hafida
International Specialization Dynamics
(Smart Innovation Set – Volume 9)

MAESSCHALCK Marc
Reflexive Governance for Research and Innovative Knowledge
(Responsible Research and Innovation Set – Volume 6)

MASSOTTE Pierre
Ethics in Social Networking and Business 1: Theory, Practice and Current Recommendations
Ethics in Social Networking and Business 2: The Future and Changing Paradigms

MASSOTTE Pierre, CORSI Patrick
Smart Decisions in Complex Systems

MEDINA Mercedes, HERRERO Mónica, URGELLÉS Alicia
Current and Emerging Issues in the Audiovisual Industry
(Diverse and Global Perspectives on Value Creation Set – Volume 1)

MICHAUD Thomas
Innovation, Between Science and Science Fiction
(Smart Innovation Set – Volume 10)

PELLÉ Sophie
Business, Innovation and Responsibility
(Responsible Research and Innovation Set – Volume 7)

SAVIGNAC Emmanuelle
The Gamification of Work: The Use of Games in the Workplace

SUGAHARA Satoshi, DAIDJ Nabyla, USHIO Sumitaka
Value Creation in Management Accounting and Strategic Management: An Integrated Approach
(Diverse and Global Perspectives on Value Creation Set –Volume 2)

UZUNIDIS Dimitri, SAULAIS Pierre
Innovation Engines: Entrepreneurs and Enterprises in a Turbulent World
(Innovation in Engineering and Technology Set – Volume 1)

2016

BARBAROUX Pierre, ATTOUR Amel, SCHENK Eric
Knowledge Management and Innovation
(Smart Innovation Set – Volume 6)

BEN BOUHENI Faten, AMMI Chantal, LEVY Aldo
Banking Governance, Performance And Risk-Taking: Conventional Banks Vs Islamic Banks

BOUTILLIER Sophie, CARRÉ Denis, LEVRATTO Nadine
Entrepreneurial Ecosystems (Smart Innovation Set – Volume 2)

BOUTILLIER Sophie, UZUNIDIS Dimitri
The Entrepreneur (Smart Innovation Set – Volume 8)

BOUVARD Patricia, SUZANNE Hervé
Collective Intelligence Development in Business

GALLAUD Delphine, LAPERCHE Blandine
Circular Economy, Industrial Ecology and Short Supply Chains (Smart Innovation Set – Volume 4)

GUERRIER Claudine
Security and Privacy in the Digital Era (Innovation and Technology Set – Volume 1)

MEGHOUAR Hicham
Corporate Takeover Targets

MONINO Jean-Louis, SEDKAOUI Soraya
Big Data, Open Data and Data Development (Smart Innovation Set – Volume 3)

MOREL Laure, LE ROUX Serge
Fab Labs: Innovative User (Smart Innovation Set – Volume 5)

PICARD Fabienne, TANGUY Corinne
Innovations and Techno-ecological Transition (Smart Innovation Set – Volume 7)

2015

CASADELLA Vanessa, LIU Zeting, DIMITRI Uzunidis
Innovation Capabilities and Economic Development in Open Economies (Smart Innovation Set – Volume 1)

CORSI Patrick, MORIN Dominique
Sequencing Apple's DNA

CORSI Patrick, NEAU Erwan
Innovation Capability Maturity Model

FAIVRE-TAVIGNOT Bénédicte
Social Business and Base of the Pyramid

GODÉ Cécile
Team Coordination in Extreme Environments

MAILLARD Pierre
Competitive Quality and Innovation

MASSOTTE Pierre, CORSI Patrick
Operationalizing Sustainability

MASSOTTE Pierre, CORSI Patrick
Sustainability Calling

2014

DUBÉ Jean, LEGROS Diègo
Spatial Econometrics Using Microdata

LESCA Humbert, LESCA Nicolas
Strategic Decisions and Weak Signals

2013

HABART-CORLOSQUET Marine, JANSSEN Jacques, MANCA Raimondo
VaR Methodology for Non-Gaussian Finance

2012

DAL PONT Jean-Pierre
Process Engineering and Industrial Management

MAILLARD Pierre
Competitive Quality Strategies

POMEROL Jean-Charles
Decision-Making and Action

SZYLAR Christian
UCITS Handbook

2011

LESCA Nicolas
Environmental Scanning and Sustainable Development

LESCA Nicolas, LESCA Humbert
Weak Signals for Strategic Intelligence: Anticipation Tool for Managers

MERCIER-LAURENT Eunika
Innovation Ecosystems

2010

SZYLAR Christian
Risk Management under UCITS III/IV

2009

COHEN Corine
Business Intelligence

ZANINETTI Jean-Marc
Sustainable Development in the USA

2008

CORSI Patrick, DULIEU Mike
The Marketing of Technology Intensive Products and Services

DZEVER Sam, JAUSSAUD Jacques, ANDREOSSO Bernadette
Evolving Corporate Structures and Cultures in Asia: Impact of Globalization

2007

AMMI Chantal
Global Consumer Behavior

2006

BOUGHZALA Imed, ERMINE Jean-Louis
Trends in Enterprise Knowledge Management

CORSI Patrick *et al.*
Innovation Engineering: the Power of Intangible Networks

Printed and bound by CPI Group (UK) Ltd, Croydon, CR0 4YY

16/09/2026

14963078-0002